T0320787

PROTECTIVE MEASUREMENT
AND QUANTUM REALITY

Protective measurements offer an intriguing method for measuring the wave function of a single quantum system. With contributions from leading physicists and philosophers of physics – including two of the original discoverers of this important method – this book explores the concept of protective measurement, investigating its broad applications and deep implications.

Addressing both physical and philosophical aspects, it covers a diverse range of topics, including the experimental possibility of protective measurements, connections with the PBR theorem, and the implications of protective measurement for understanding the nature of quantum reality. Including a clear and concise introduction to standard quantum mechanics, conventional measurement, and the fundamentals of protective measurement, this is a valuable resource for graduate students and researchers interested in the conceptual foundations of quantum mechanics.

SHAN GAO is an Associate Professor at the Institute for the History of Natural Sciences, Chinese Academy of Sciences. His research focuses on the foundations of quantum mechanics and the history of modern physics.

PROTECTIVE MEASUREMENT AND QUANTUM REALITY

Towards a New Understanding of Quantum Mechanics

SHAN GAO

Chinese Academy of Sciences

CAMBRIDGE
UNIVERSITY PRESS

University Printing House, Cambridge CB2 8BS, United Kingdom

Cambridge University Press is part of the University of Cambridge.

It furthers the University's mission by disseminating knowledge in the pursuit of education, learning and research at the highest international levels of excellence.

www.cambridge.org
Information on this title: www.cambridge.org/9781107069633

© Cambridge University Press 2014

This publication is in copyright. Subject to statutory exception and to the provisions of relevant collective licensing agreements, no reproduction of any part may take place without the written permission of Cambridge University Press.

First published 2014

A catalogue record for this publication is available from the British Library

Library of Congress Cataloguing in Publication data
Protective measurement and quantum reality : toward a new understanding of quantum mechanics / edited by Shan Gao, Chinese Academy of Science.
pages   cm.
Includes bibliographical references and index.
ISBN 978-1-107-06963-3 (Hardback)
1. Quantum theory.   2. Physical measurements.
I. Gao, Shan, 1970 or 1971- editor, author.
QC174.13.P76 2014
530.12–dc23   2014021791

ISBN 978-1-107-06963-3 Hardback

Cambridge University Press has no responsibility for the persistence or accuracy of URLs for external or third-party internet websites referred to in this publication, and does not guarantee that any content on such websites is, or will remain, accurate or appropriate.

To my parents

Contents

Contributors

Yakir Aharonov, Tel Aviv University and Chapman University

Gennaro Auletta, University of Cassino

Tangereen V. B. Claringbold, University of Portland

Eliahu Cohen, Tel Aviv University

Michael Dickson, University of South Carolina

Lajos Diósi, Wigner Research Center for Physics

Mauro Dorato, University of Rome Three

Aurélien Drezet, CNRS-University Joseph Fourier

Robert Flack, University College London

Shan Gao, Chinese Academy of Sciences

Guy Hetzroni, Hebrew University of Jerusalem

Basil J. Hiley, University of London

Peter Holland, University of Oxford

Vincent Lam, University of Lausanne

Federico Laudisa, University of Milan-Bicocca

Peter J. Lewis, University of Miami

Daniel Rohrlich, Ben-Gurion University of the Negev

Maximilian Schlosshauer, University of Portland

Lev Vaidman, Tel Aviv University

Preface

In 1993, Yakir Aharonov, Lev Vaidman and Jeeva Anandan discovered an important new method of measurement in quantum mechanics, the so-called protective measurement. Distinct from conventional measurements, protective measurement is a method for measuring the expectation value of an observable on a single quantum system. By a series of protective measurements, one can even measure the wave function of a single quantum system. In this way, theoretical analysis of protective measurement may lead to a new and deeper understanding of quantum mechanics. Moreover, its experimental realization may also be useful for quantum information technology.

This book is an anthology celebrating the 20th anniversary of the discovery of protective measurement. It begins with a clear and concise introduction to standard quantum mechanics, conventional measurement and protective measurement, and contains fourteen original essays written by physicists and philosophers of physics, including Yakir Aharonov and Lev Vaidman, the two discoverers. The topics include the fundamentals of protective measurement, its meaning and applications, and current views on the importance and implications of protective measurement. The book is accessible to graduate students in physics and chemistry. It will be of value to students and researchers with an interest in the meaning of quantum theory and especially to physicists and philosophers working on the foundations of quantum mechanics.

When I contacted potential contributors to this anthology, one of them replied, "Protective measurements are something I know nothing about." Indeed, as one referee of this book also admitted, although protective measurement has attracted attention over the last 20 years and has raised many interesting questions, it is still an under-studied aspect of quantum mechanics. In recent years the associated field of weak measurement has seen significant increased activity, and the latest Pusey–Barrett–Rudolph theorem has also caused many people to revisit the question of the reality of the wave function. Can protective measurement, like weak measurement,

be performed in laboratories in the near future? Do protective measurements anticipate the Pusey–Barrett–Rudolph theorem? What, if any, are the implications of
protective measurements for the ontological meaning of the wave function and the
nature of quantum reality? I hope this anthology will arouse more researchers'
interest in protective measurement and its implications, and further open up a new
line of research in the foundations of quantum mechanics.

I wish to express my warm thanks to Baichun Zhang, Yidong Liu and Miao
Tian for helpful discussions, which inspired me to take up the project of editing an
anthology about protective measurement and relevant topics. I am grateful to Yakir
Aharonov and Lev Vaidman for their support for the project. I thank all contributors
for taking the time to write these new essays in the anthology. I also thank Simon
Capelin of Cambridge University Press for his kind support as I worked on this
project, and the three referees who gave helpful suggestions on how the work could
best serve its targeted audience. Finally, I am deeply indebted to my wife Huixia
and my daughter Ruiqi for their unflagging love and support.

Shan Gao
Beijing, 2013

Acknowledgements

The editor and publisher gratefully acknowledge the permission granted to reproduce copyright texts and figures on the following pages:

p.71: Figure 6.1. Reprinted with permission from Kocsis, S., Braverman, B., Ravets, S., *et al.* (2011a). Observing the average trajectories of single photons in a two-slit interferometer, *Science*. **332**, 1170–1173;

p.74: Figure 6.4. Reprinted with permission from Duck, I. M., Stevenson, P. M. and Sudarshan, E. C. G. (1989). The sense in which a "weak measurement" of a spin-1/2 particle's spin component yields a value 100, Phys. Rev. D, **40**, 2112–2117;

p.77: Figure 6.5. Reprinted with permission from Ritchie, N. W., Story, J. G. and Hulet, R. G. (1991). Realization of a measurement of a weak value, *Phys. Rev. Lett.*, **66**, 1107–1110;

p.145: Reprinted with permission from Frost, R. (1916). *The Road Not Taken and Other Poems*. New York: Dover. (1993 reissue of *Mountain Interval*. New York: Henry Holt & Co. 1916).

1

Protective measurement: an introduction

SHAN GAO

Protective measurement, in the language of standard quantum mechanics, is a method to measure the *expectation value* of an observable on a *single* quantum system (Aharonov and Vaidman, 1993; Aharonov, Anandan and Vaidman, 1993). For a conventional impulsive measurement, the state of the measured system is strongly entangled with the state of the measuring device during the measurement, and the measurement result is one of the eigenvalues of the measured observable. By contrast, during a protective measurement, the measured state is protected by an appropriate procedure so that it neither changes nor becomes entangled with the state of the measuring device appreciably. In this way, such protective measurements can measure the expectation values of observables on a single quantum system, and in particular, the wave function of the system can also be measured as expectation values of certain observables. It is expected that protective measurements can be performed in the near future with the rapid development of weak measurement technologies (e.g. Kocsis et al., 2011; Lundeen et al., 2011). In this chapter, we will give a clear introduction to protective measurement in quantum mechanics.

1.1 Standard quantum mechanics and impulsive measurement

The standard formulation of quantum mechanics, which was first developed by Dirac (1930) and von Neumann (1955), is based on the following basic principles.

1 Physical states

The state of a physical system is represented by a normalized wave function or unit vector $|\psi(t)\rangle$ in a Hilbert space.[1] The Hilbert space is complete in the sense that every possible physical state can be represented by a state vector in the space.

---

[1] The Hilbert space is a complete vector space with scalar product. The common notion of state includes both proper vectors normalizable to unity in the Hilbert space and so-called improper vectors normalizable only to the Dirac delta functions.

2 Physical properties

Every measurable property or observable of a physical system is represented by a Hermitian operator on the Hilbert space associated with the system. A physical system has a determinate value for an observable if and only if it is in an eigenstate of the observable (this is often called the eigenvalue–eigenstate link).

3 Composition rule

The Hilbert space associated with a composite system is the tensor product of the Hilbert spaces associated with the systems of which it is composed. Similarly, the Hilbert space associated with independent properties is the tensor product of the Hilbert spaces associated with each property.

4 Evolution law

(1) Linear evolution

The state of a physical system $|\psi(t)\rangle$ obeys the linear Schrödinger equation $i\hbar\partial|\psi(t)\rangle/\partial t = H|\psi(t)\rangle$ (when it is not measured), where $\hbar$ is Planck's constant divided by 2π, and H is the Hamiltonian operator that depends on the energy properties of the system.

(2) Non-linear collapse evolution

If a physical system is in a state $|\psi\rangle = \sum_i c_i|a_i\rangle$, where $|a_i\rangle$ is the eigenstate of an observable A with eigenvalue a_i, then an (impulsive) measurement of the observable A will instantaneously and randomly collapse the state into one of the eigenstates $|a_i\rangle$ with probability $|c_i|^2$. This is called the collapse postulate, and the non-linear stochastic process is called the reduction of the state vector or the collapse of the wave function.

The link between the mathematical formalism and experiments is provided by the Born rule. It says that the probability of the above measurement of the observable A yielding the result a_i is $|c_i|^2$. For a continuous property such as position x, the probability of obtaining a measurement result between x and $x + dx$ is $|\langle x|\psi\rangle|^2 dx$. Note that the Born rule can be derived from the collapse postulate by resorting to the eigenvalue–eigenstate link, but it does not necessarily depend on the postulate.

The conventional impulsive measurements can be formulated as follows. According to the standard von Neumann procedure, measuring an observable A in a quantum state $|\psi\rangle$ involves an interaction Hamiltonian

$$H_{\mathrm{I}} = g(t)PA \tag{1.1}$$

coupling the measured system to an appropriate measuring device, where P is the conjugate momentum of the pointer variable X of the device. The time-dependent coupling strength $g(t)$ is a smooth function normalized to $\int dt g(t) = 1$ during the measurement interval τ, and $g(0) = g(\tau) = 0$. The initial state of the pointer at

$t = 0$ is supposed to be a Gaussian wave packet of eigenstates of X with width w_0, centered around the eigenvalue x_0, which is denoted by $|\phi(x_0)\rangle$.

For an impulsive measurement, the interaction H_I is of very short duration and so strong that it dominates the rest of the Hamiltonian (i.e. the effect of the free Hamiltonians of the measuring device and the measured system can be neglected). Then the state of the combined system at the end of the interaction can be written as

$$|t = \tau\rangle = e^{-\frac{i}{\hbar}PA} |\psi\rangle |\phi(x_0)\rangle. \tag{1.2}$$

By expanding $|\psi\rangle$ in the eigenstates of A, $|a_i\rangle$, we obtain

$$|t = \tau\rangle = \sum_i e^{-\frac{i}{\hbar}Pa_i} c_i |a_i\rangle |\phi(x_0)\rangle, \tag{1.3}$$

where c_i are the expansion coefficients. The exponential term shifts the center of the pointer by a_i:

$$|t = \tau\rangle = \sum_i c_i |a_i\rangle |\phi(x_0 + a_i)\rangle. \tag{1.4}$$

This is an entangled state, where the eigenstates of A with eigenvalues a_i get correlated to measuring device states in which the pointer is shifted by these values a_i. Then, by the collapse postulate, the state will instantaneously and randomly collapse into one of its branches $|a_i\rangle |\phi(x_0 + a_i)\rangle$ with probability $|c_i|^2$. This means that the measurement result can only be one of the eigenvalues of the measured observable A, say a_i, with a certain probability $|c_i|^2$. The expectation value of A is then obtained as the statistical average of eigenvalues for an ensemble of identically prepared systems, namely $\langle A \rangle = \sum_i |c_i|^2 a_i$.

1.2 Weak measurement

Impulsive measurements are only one kind of quantum measurement, for which the coupling between the measured system and the measuring device is very strong, and the results are only the eigenvalues of an observable. We can also perform other kinds of measurement by adjusting the coupling strength. An interesting example is weak measurement (Aharonov, Albert and Vaidman, 1988; Aharonov and Vaidman, 1990), for which the measurement result is the expectation value of the measured observable.

A weak measurement is a standard measuring procedure with weakened coupling. Like impulsive measurements, the interaction Hamiltonian is also given by (1.1) for a weak measurement. The weakness of the interaction is achieved by preparing the initial state of the measuring device in such a way that the conjugate momentum of the pointer variable is localized around zero with small uncertainty, and thus the interaction Hamiltonian (1.1) is small. As a simple example, let the initial state of the pointer in position space be:

$$\langle x|\phi\rangle = (w_0^2\pi)^{-1/4}e^{-x^2/2w_0^2}. \tag{1.5}$$

The corresponding initial probability distribution is

$$P_i(x) = (w_0^2\pi)^{-1/2}e^{-x^2/w_0^2}. \tag{1.6}$$

Expanding the initial state of the system $|\psi\rangle$ in the eigenstates $|a_i\rangle$ of the measured observable A, $|\psi\rangle = \sum_i c_i|a_i\rangle$, then after the interaction (1.1) the state of the system and the measuring device is:

$$|t = \tau\rangle = (w_0^2\pi)^{-1/4}\sum_i c_i|a_i\rangle e^{-(x-a_i)^2/2w_0^2}. \tag{1.7}$$

The probability distribution of the pointer variable corresponding to the final state (1.7) is:

$$P_f(x) = (w_0^2\pi)^{-1/2}\sum_i |c_i|^2 e^{-(x-a_i)^2/w_0^2}. \tag{1.8}$$

In the case of an impulsive measurement, this is a weighted sum of the initial probability distribution localized around various eigenvalues a_i. Therefore, the reading of the pointer variable at the end of the measurement always yields a value close to one of the eigenvalues. By contrast, the limit of weak measurement corresponds to $w_0 \gg a_i$ for all eigenvalues a_i. Then we can perform the Taylor expansion of the sum (1.8) around $x = 0$ up to first order and rewrite the final probability distribution of the pointer variable in the following way:

$$P_f(x) \approx (w_0^2\pi)^{-1/2}\sum_i |c_i|^2[1 - (x - a_i)^2/w_0^2] \approx (w_0^2\pi)^{-1/2}e^{-(x-\sum_i |c_i|^2 a_i)^2/w_0^2}. \tag{1.9}$$

This is the initial probability distribution shifted by the value $\sum_i |c_i|^2 a_i$ (Aharonov and Vaidman, 2008). It indicates that the result of the weak measurement is the expectation value of the measured observable in the measured state:

$$\langle A\rangle \equiv \langle\psi|A|\psi\rangle = \sum_i |c_i|^2 a_i. \tag{1.10}$$

Certainly, since the width of the pointer wave packet is much greater than the shift of the center of the pointer, namely $w_0 \gg \langle A\rangle$, the above weak measurement of a single system is very imprecise. However, by performing the weak measurement on an ensemble of N identically prepared systems the precision can be improved by a factor $\sqrt{N}$. This scheme of weak measurement has been realized and proved useful in quantum optical experiments (see, e.g. Kocsis et al., 2011; Lundeen et al., 2011).

Although weak measurements, like conventional impulsive measurements, also need to measure an ensemble of identically prepared quantum systems, they are conceptually different. For impulsive measurements, every identically prepared system in the ensemble shifts the pointer of the measuring device by one of the eigenvalues of the measured observable. By contrast, for weak measurements, every identically prepared system in the ensemble shifts the pointer of the measuring device directly by the expectation value of the measured observable.

1.3 Protective measurement

Protective measurements are improved methods of weak measurements in the sense that they can measure the expectation values of observables on a single quantum system (Aharonov and Vaidman, 1993; Aharonov, Anandan and Vaidman, 1993). For an impulsive measurement, if the measured system, prior to the measurement of an observable A, is not in an eigenstate of A, then its state will be invariably entangled with the state of the device due to the interaction. A protective measurement differs from an impulsive measurement (as well as from a weak measurement) in that the measured state is protected from being entangled and changed appreciably when the measurement is being made. A universal protection scheme is via the quantum Zeno effect. Let's see how this can be done.

1.3.1 Measurements with artificial protection

Let $|\psi\rangle$ be an arbitrary known state of a single quantum system at a given instant $t = 0$. To protect this state from being changed, we make projective measurements of an observable $P(t)$, for which $|\psi\rangle$ is a non-degenerate eigenstate, a large number of times which are dense in the measurement interval $[0, \tau]$ (Aharonov, Anandan and Vaidman, 1993). For example, $P(t)$ is measured in $[0, \tau]$ at times $t_n = (n/N)\tau, n = 1, 2, ..., N$, where N is an arbitrarily large number. At the same time, we make an impulsive measurement of observable A in the interval $[0, \tau]$, which is described by the interaction Hamiltonian (1.1). The initial state of the pointer is supposed to be a Gaussian wave packet of width w_0 centered at initial position x_0, denoted by $|\phi(x_0)\rangle$.

Then the branch of the state of the combined system after τ, in which each projective measurement of $P(t_n)$ results in the state of the measured system being in $|\psi\rangle$, is given by

$$
\begin{aligned}
|t = \tau\rangle &= |\psi\rangle \langle\psi| e^{-\frac{i}{\hbar}\frac{\tau}{N}H(t_N)} ... |\psi\rangle \langle\psi| e^{-\frac{i}{\hbar}\frac{\tau}{N}H(t_2)} |\psi\rangle \langle\psi| \\
&\qquad \times e^{-\frac{i}{\hbar}\frac{\tau}{N}H(t_1)} |\psi\rangle |\phi(x_0)\rangle \\
&= |\psi\rangle \langle\psi| e^{-\frac{i}{\hbar}\frac{\tau}{N}g(t_N)PA} ... |\psi\rangle \langle\psi| e^{-\frac{i}{\hbar}\frac{\tau}{N}g(t_2)PA} |\psi\rangle \langle\psi| \\
&\qquad \times e^{-\frac{i}{\hbar}\frac{\tau}{N}g(t_1)PA} |\psi\rangle |\phi(x_0)\rangle.
\end{aligned}
\tag{1.11}
$$

Thus in the limit of $N \to \infty$, we have

$$|t = \tau\rangle = |\psi\rangle \, e^{-\frac{i}{\hbar} \int_0^\tau g(t)\langle\psi|A|\psi\rangle P dt} \, |\phi(x_0)\rangle = |\psi\rangle \, |\phi(x_0 + \langle A\rangle)\rangle. \qquad (1.12)$$

Since the total probability of other branches is proportional to τ^2/N to first order of N, the above state will be the state of the combined system after τ when $N \to \infty$. This demonstrates that for an arbitrary but known state of a quantum system at a given instant, we can protect the state from being changed via the quantum Zeno effect by frequent projective measurements, and an independent measurement of an observable A, which is made at the same time, yields the expectation value of the observable in the measured state.

By a conventional impulsive measurement on a single quantum system, one obtains one of the eigenvalues of the measured observable, and the expectation value of the observable can only be obtained as the statistical average of eigenvalues for an ensemble of identically prepared systems. Thus it seems surprising that a protective measurement can yield the expectation value of the measured observable directly from a single quantum system. In fact, the appearance of expectation values as measurement results is quite natural when the measured state is not changed and the entanglement during conventional measurements does not take place as for protective measurements (Aharonov, Anandan and Vaidman, 1993). In this case, the evolution of the combining state is

$$|\psi(0)\rangle \, |\phi(0)\rangle \to |\psi(t)\rangle \, |\phi(t)\rangle, t > 0, \qquad (1.13)$$

where $|\psi(t)\rangle$ is the same as $|\psi(0)\rangle$ up to a phase factor during the measurement interval $[0, \tau]$. The interaction Hamiltonian is still given by (1.1). Then, by Ehrenfest's theorem we have

$$\frac{d}{dt}\langle\psi(t)\phi(t)|X|\psi(t)\phi(t)\rangle = g(t)\langle\psi(0)|A|\psi(0)\rangle, \qquad (1.14)$$

where X is the pointer variable. This further leads to

$$\langle\phi(\tau)|X|\phi(\tau)\rangle - \langle\phi(0)|X|\phi(0)\rangle = \langle\psi(0)|A|\psi(0)\rangle. \qquad (1.15)$$

This means that the shift of the center of the pointer of the device gives the expectation value of the measured observable in the measured state.

1.3.2 Measurements with natural protection

In some special cases, the universal protection procedure via the quantum Zeno effect is not necessary, and the system's Hamiltonian can help protect its state from changing when the measurement interaction is weak and adiabatic. For example, for a quantum system in a discrete non-degenerate energy eigenstate, the system itself supplies the protection of the state due to energy conservation. By the adiabatic theorem, the adiabatic interaction during the measurement ensures that the measured system cannot make a transition from one discrete energy eigenstate to another. Moreover, according to first-order perturbation theory, for any given value of P, the energy of the measured energy eigenstate shifts by an infinitesimal amount: $\delta E = \langle H_I \rangle = g(t)P\langle A \rangle$, and the corresponding time evolution $\mathrm{e}^{-iP\langle A \rangle/\hbar}$ then shifts the pointer by the expectation value $\langle A \rangle$. For degenerate energy eigenstates, we may not use the universal protection procedure either. The simplest way is to add a protective potential to change the energies of other states and lift the degeneracy. Then the measured state remains unchanged, but is now protected by energy conservation like non-degenerate energy eigenstates.

As a simple example, we consider a quantum system in a discrete non-degenerate energy eigenstate $|E_n\rangle$. In this case, the system itself supplies the protection of the state and no artificial protection is needed. The interaction Hamiltonian for a protective measurement of an observable A in this state is also given by (1.1) as for conventional impulsive measurements. But differently from impulsive measurements, for which the interaction is very strong and almost instantaneous, the protective measurements make use of the opposite limit where the interaction of the measuring device with the system is weak and adiabatic, and thus the free Hamiltonians cannot be neglected. Let the total Hamiltonian of the combined system be

$$H = H_S + H_D + H_I, \tag{1.16}$$

where H_S and H_D are the free Hamiltonians of the measured system and the measuring device, respectively, and $H_I = g(t)PA$ is the interaction Hamiltonian. As before, we suppose the time-dependent coupling strength $g(t)$ is a smooth function normalized to $\int \mathrm{d}t g(t) = 1$ in the measurement interval $[0, T]$, and $g(0) = g(T) = 0$, and the initial state of the pointer is a Gaussian wave packet of width w_0 centered at initial position x_0, denoted by $|\phi(x_0)\rangle$.

The state of the combined system after T is then given by

$$|t = T\rangle = \mathrm{e}^{-\frac{i}{\hbar}\int_0^T H(t)\mathrm{d}t} |E_n\rangle |\phi(x_0)\rangle . \tag{1.17}$$

By ignoring the switching on and switching off processes, the full Hamiltonian (with $g(t) = 1/T$) is time-independent and no time-ordering is needed.[2] Then we obtain

$$|t = T\rangle = \mathrm{e}^{-\frac{i}{\hbar}HT} |E_n\rangle |\phi(x_0)\rangle , \tag{1.18}$$

---

[2] The change in the total Hamiltonian during these processes is smaller than PA/T, and thus the approximate treatment given below is valid. For a more strict analysis see Dass and Qureshi (1999).

where $H = H_S + H_D + PA/T$. We further expand $|\phi(x_0)\rangle$ in the eigenstate of H_D, $\left|E_j^d\right\rangle$, and write

$$|t = T\rangle = e^{-\frac{i}{\hbar}HT} \sum_j c_j |E_n\rangle |E_j^d\rangle. \tag{1.19}$$

Let the exact eigenstates of H be $|\Psi_{k,m}\rangle$ and the corresponding eigenvalues be $E(k, m)$; we have

$$|t = T\rangle = \sum_j c_j \sum_{k,m} e^{-\frac{i}{\hbar}E(k,m)T} \langle\Psi_{k,m}|E_n, E_j^d\rangle |\Psi_{k,m}\rangle. \tag{1.20}$$

Since the interaction is very weak, the total Hamiltonian H can be thought of as $H_0 = H_S + H_D$ perturbed by PA/T. Using the fact that PA/T is a small perturbation and that the eigenstates of H_0 are of the form $|E_k\rangle |E_m^d\rangle$, the perturbation theory gives

$$\left|\Psi_{k,m}\right\rangle = |E_k\rangle |E_m^d\rangle + O(1/T),$$

$$E(k, m) = E_k + E_m^d + \frac{1}{T}\langle A\rangle_k \langle P\rangle_m + O(1/T^2). \tag{1.21}$$

Note that it is a necessary condition for (1.21) to hold that $|E_k\rangle$ is a non-degenerate eigenstate of H_S. Substituting (1.21) in (1.20) and taking the limit $T \to \infty$ yields

$$|t = T\rangle_{T\to\infty} = \sum_j e^{-\frac{i}{\hbar}(E_n T + E_j^d T + \langle A\rangle_n \langle P\rangle_j)} c_j |E_n\rangle |E_j^d\rangle. \tag{1.22}$$

For the case where P commutes with the free Hamiltonian of the device,[3] i.e., $[P, H_D] = 0$, the eigenstates $|E_j^d\rangle$ of H_D are also the eigenstates of P, and thus the above equation can be rewritten as

$$|t = T\rangle_{T\to\infty} = e^{-\frac{i}{\hbar}E_n T - \frac{i}{\hbar}H_D T - \frac{i}{\hbar}\langle A\rangle_n P} |E_n\rangle |\phi(x_0)\rangle. \tag{1.23}$$

It can be seen that the third term in the exponent will shift the center of the pointer by an amount $\langle A\rangle_n$:

$$|t = T\rangle_{T\to\infty} = e^{-\frac{i}{\hbar}E_n T - \frac{i}{\hbar}H_D T} |E_n\rangle |\phi(x_0 + \langle A\rangle_n)\rangle. \tag{1.24}$$

This indicates that the result of the protective measurement is the expectation value of the measured observable in the measured state, and moreover, the measured state is not changed by the protective measurement (except for an overall phase factor).

It is worth noting that since the position variable of the pointer does not commute with its free Hamiltonian, the pointer wave packet will spread during the measurement interval. For example, the kinematic energy term $P^2/2M$ in the free Hamiltonian of the pointer will spread the wave packet without shifting the center, and the

---

[3] For the derivation for the case $[P, H_D] \neq 0$ see Dass and Qureshi (1999).

width of the wave packet at the end of interaction will be $w(T) = [\frac{1}{2}(w_0^2 + \frac{T^2}{M^2 w_0^2})]^{\frac{1}{2}}$ (Dass and Qureshi, 1999). However, the spreading of the pointer wave packet can be made as small as possible by increasing the mass M of the pointer, and thus it will not interfere with resolving the shift of the center of the pointer in principle.

1.3.3 Measurements of the wave function of a single system

Since the wave function can be reconstructed from the expectation values of a sufficient number of observables, the wave function of a single quantum system can be measured by a series of protective measurements. Let the explicit form of the measured state at a given instant t be $\psi(x)$, and the measured observable A be (normalized) projection operators on small spatial regions V_n having volume v_n:

$$A = \begin{cases} \frac{1}{v_n}, & \text{if } x \in V_n, \\ 0, & \text{if } x \notin V_n. \end{cases} \tag{1.25}$$

A protective measurement of A then yields

$$\langle A \rangle = \frac{1}{v_n} \int_{V_n} |\psi(x)|^2 \mathrm{d}v, \tag{1.26}$$

which is the average of the density $\rho(x) = |\psi(x)|^2$ over the small region V_n. Similarly, we can measure another observable $B = \frac{\hbar}{2mi}(A\nabla + \nabla A)$. The measurement yields

$$\langle B \rangle = \frac{1}{v_n} \int_{V_n} \frac{\hbar}{2mi}(\psi^* \nabla \psi - \psi \nabla \psi^*) \mathrm{d}v = \frac{1}{v_n} \int_{V_n} j(x) \mathrm{d}v. \tag{1.27}$$

This is the average value of the flux density $j(x)$ in the region V_n. Then when $v_n \to 0$ and after performing measurements in sufficiently many regions V_n we can measure $\rho(x)$ and $j(x)$ everywhere in space. Since the wave function $\psi(x, t)$ can be uniquely expressed by $\rho(x, t)$ and $j(x, t)$ (except for an overall phase factor), the whole wave function of the measured system at a given instant can be measured by protective measurements.

1.4 Further discussion

Protective measurement is a surprising measuring method, by which one can measure the *expectation value* of an observable on a *single* quantum system, even if the system is not in an eigenstate of the measured observable. This remarkable feature makes protective measurements quite distinct from conventional impulsive measurements. It is unsurprising that there appeared numerous objections to the validity and meaning of protective measurements (see, e.g. Unruh, 1994; Rovelli, 1994; Ghose and Home, 1995; Uffink, 1999, 2013). Although misunderstandings have been clarified (Aharonov, Anandan and Vaidman, 1996; Dass and Qureshi, 1999; Vaidman, 2009; Gao, 2013), it is still debatable whether protective measurement has important implications for our understanding of quantum mechanics, especially for the ontological status of the wave function. In the following, we will emphasize three key points that may help us understand protective measurement, and briefly review the current state of debate on its possible implications.

First of all, a single quantum system being in an arbitrary known state can be protectively measured in principle. The state of the system being protected to be unchanged permits the state as well as the expectation values of observables in the state to be measurable. In this sense, protective measurements are not a special kind of quantum measurement, but the *very* way to measure the actual state of a quantum system at a given instant. By comparison, a non-protective measurement such as an impulsive measurement will change the measured state, and the resulting measurement outcome (i.e. one of the eigenvalues of the measured observable) does not reflect the original state of the measured system. Besides, when a quantum system interacts with another system under non-protective conditions, its state also evolves in time, and thus the expectation values of observables do not manifest themselves explicitly in the interaction either. For example, the interaction between two charged quantum systems is not directly dependent on the expectation values of their charges, but described by the potential terms in the Schrödinger equation (see Chapter 15).

Next, a realistic protective measurement can never be performed on a single quantum system with absolute certainty. For example, for a realistic protective measurement of an observable A on a non-degenerate energy eigenstate whose measurement interval T is finite, there is always a tiny probability proportional to $1/T^2$ of obtaining a different result $\langle A \rangle_\perp$, where $\perp$ refers to a normalized state in the subspace normal to the measured state as picked out by first-order perturbation theory. However, this effect can be made arbitrarily small when the measurement interval T is arbitrarily long. In this sense, an ideal protective measurement can measure the expectation values of observables on a single quantum system with certainty *in principle*.

Thirdly, we stress that the validity of the scheme of protective measurements does not rely on the standard von Neumann formulation of measurements. In the above formulation of protective measurement, the measuring system can be a microscopic system such as an electron, and the shift of the center of the wave packet of the measuring system is only determined by the Schrödinger equation. Since the state of the measured system is not changed during the protective measurement, a large number of identically prepared measuring systems can be used to protectively measure the original measured system, and the centers of their wave packets have the same shift. Then the shift can be read out by conventional impulsive measurements of the ensemble of these measuring systems, for which the probability distribution of the results satisfies the Born rule. In summary, the scheme of protective measurement is only based on the Schrödinger equation and the Born rule, and especially, it is independent of whether wave function collapse is real or apparent.

Finally, we will briefly review the current state of debate on the implications of protective measurement for the reality of the wave function (see also Chapter 2). According to the standard view, the expectation values of observables are not the physical properties of a single system, but the statistical properties of an ensemble of identical systems. This seems reasonable if there exist only conventional impulsive measurements. An impulsive measurement obtains one of the eigenvalues of the measured observable, and the expectation value can only be defined as a statistical average of the eigenvalues for an ensemble of identical systems. However, there exist other kinds of quantum measurement, and in particular, protective measurements can measure the expectation values of observables from a single system. Therefore, it seems that the expectation values of observables should be taken as the physical properties of a single quantum system. Moreover, since the wave function can be reconstructed from the expectation values of a sufficient number of observables, this result will further imply that the wave function represents the physical state of a single quantum system.

Several authors, including the discoverers of protective measurements, have given a similar argument as above (Aharonov and Vaidman, 1993; Aharonov, Anandan and Vaidman, 1993; Anandan, 1993; Dickson, 1995). According to Aharonov and Vaidman (1993), the existence of protective measurements provides a strong argument for associating physical reality with the wave function of a single system, and challenges the standard view that the wave function has physical meaning only for an ensemble of identical systems. Anandan (1993) further argued that protective measurement refutes an argument of Einstein in favor of the ensemble interpretation of quantum mechanics. In addition, according to Dickson (1995), protective measurement provides a reply to scientific empiricism about quantum mechanics (see also Chapter 8).

However, these analyses have been neglected by most researchers, and they are also subject to some objections (Unruh, 1994; Dass and Qureshi, 1999; Lewis, Chapter 7; Schlosshauer and Claringbold, Chapter 13). There are mainly two objections to the above implications of protective measurements. The first one claims that since an unknown state of a single system cannot be protectively measured, protective measurements do not have implications for the ontological status of the wave function (Unruh, 1994). According to the second objection, that a realistic protective measurement can never be performed on a single quantum system with absolute certainty, an ontological status for the wave function is precluded

(Dass and Qureshi, 1999; Schlosshauer and Claringbold, Chapter 7). These objections will be answered in this volume by several authors (Vaidman, Chapter 2; Hetzroni and Rohrlich, Chapter 10; Gao, Chapter 15). According to their analysis, protective measurements will help unveil the reality and meaning of the wave function, and lead to a new and deeper understanding of quantum mechanics.

References

Aharonov, Y. and Vaidman, L. (1990). Properties of a quantum system during the time interval between two measurements. *Phys. Rev. A* **41**, 11.

Aharonov, Y. and Vaidman, L. (1993). Measurement of the Schrödinger wave of a single particle. *Phys. Lett. A* **178**, 38.

Aharonov, Y. and Vaidman, L. (2008). The two-state vector formalism: an updated review. *Lect. Notes Phys.* **734**, 399.

Aharonov, Y., Albert, D. Z. and Vaidman, L. (1988). How the result of a measurement of a component of the spin of a spin-1/2 particle can turn out to be 100. *Phys. Rev. Lett.* **60**, 1351.

Aharonov, Y., Anandan, J. and Vaidman, L. (1993). Meaning of the wave function. *Phys. Rev. A* **47**, 4616.

Aharonov, Y., Anandan, J. and Vaidman, L. (1996). The meaning of protective measurements. *Found. Phys.* **26**, 117.

Anandan, J. (1993). Protective measurement and quantum reality. *Found. Phys. Lett.* **6**, 503–532.

Dass, N. D. H. and Qureshi, T. (1999). Critique of protective measurements. *Phys. Rev. A* **59**, 2590.

Dickson, M. (1995). An empirical reply to empiricism: protective measurement opens the door for quantum realism. *Philosophy of Science* **62**, 122.

Dirac, P. A. M. (1930). *The Principles of Quantum Mechanics*. Oxford: Clarendon Press.

Gao, S. (2013). On Uffink's criticism of protective measurements. *Studies in History and Philosophy of Modern Physics* **44**, 513–518.

Ghose, P. and Home, D. (1995). An analysis of the Aharonov–Anandan–Vaidman model. *Found. Phys.* **25**, 1105.

Kocsis, S., Braverman, B., Ravets, S. et al. (2011). Observing the average trajectories of single photons in a two-slit interferometer. *Science* **332**, 1170–1173.

Lundeen, J. S., Sutherland, B., Patel, A., Stewart, C. and Bamber, C. (2011). Direct measurement of the quantum wavefunction. *Nature* **474**, 188–191.

Rovelli, C. (1994). Comment on "Meaning of the wave function". *Phys. Rev. A* **50**, 2788.

Uffink, J. (1999). How to protect the interpretation of the wave function against protective measurements. *Phys. Rev. A* **60**, 3474–3481.

Uffink, J. (2013). Reply to Gao's "On Uffink's criticism of protective measurements". *Studies in History and Philosophy of Modern Physics* **44**, 519–523.

Unruh, W. G. (1994). Reality and measurement of the wave function. *Phys. Rev. A* **50**, 882.

Vaidman, L. (2009). Protective measurements, in Greenberger, D., Hentschel, K. and Weinert, F. (eds.), *Compendium of Quantum Physics: Concepts, Experiments, History and Philosophy*. Berlin: Springer-Verlag, pp. 505–507.

von Neumann, J. (1955). *Mathematical Foundations of Quantum Mechanics*. Princeton: Princeton University Press. (Translated by R. Beyer from *Mathematische Grundlagen der Quantenmechanik*. Berlin: Springer, 1932.)

Part I

Fundamentals and applications

2

Protective measurement of the wave function of a single system

LEV VAIDMAN

My view on the meaning of the quantum wave function and its connection to protective measurements is described. The wave function and only the wave function is the ontology of the quantum theory. Protective measurements support this view although they do not provide a decisive proof. A brief review of the discovery and the criticism of protective measurement is presented. Protective measurements with postselection are discussed.

2.1 Introduction

In the first graduate course of quantum mechanics I remember asking the question: "Can we consider the wave function as a description of a single quantum system?" I got no answer. Twelve years later, in South Carolina, after I completed my Ph.D. studies at Tel Aviv University under the supervision of Yakir Aharonov in which we developed the theory of weak measurements [1], I asked Aharonov: Can we use weak measurement to observe the wave function of a single particle?

At that time I had already become a strong believer in the many-worlds interpretation (MWI) of quantum mechanics [2] and had no doubt that a single system *is* described by the wave function. Yakir Aharonov never shared with me the belief in the MWI. When we realized that using what is called now *protective measurement*, we can, under certain conditions, observe the wave function of a single quantum system, he was really excited by the result. At 1992 I was invited to a conference on the Foundations of Quantum Mechanics in Japan where I presented this result: "The Schrödinger wave is observable after all!"[3]. Then I went home to Tel Aviv where I finished writing a letter which received mixed reviews in *Phys. Rev. Lett.*, while Jeeva Anandan, working on the topic with Aharonov in South Carolina, wrote a paper accepted in *Phys. Rev. A* [4]. After acceptance of the PRA paper it was hard to fight the referees in PRL, but PLA accepted it immediately [5].

15

I do not think that protective measurements provide a decisive answer to my question in the graduate school. I came to believe that a single quantum particle is completely described by its Schrödinger wave before understanding protective measurements. And after the publication of our work on protective measurement, many physicists still view it as an open question. This manifests in the enormous interest to the Pusey, Barrett, and Rudolph (PBR) paper [6] entitled "On the reality of the quantum state" which puts strong constraints on the ensemble interpretation. Also, one of the "most read" *Nature* papers is "Direct measurement of the quantum wavefunction" [7].

The method of protective measurement provides a plausibility argument: it is more natural to attribute the wave function to a single particle when there is a procedure to observe it. It is not a decisive argument because there are some limitations on the measurability of the wave function of a single particle. The quantum state has to be "protected", so it is possible to argue that what is measured is not the wave function, but the "protection procedure". Still, I can argue that it is better to say that we measure the quantum state and not its "protection" because there are many different protections for the same wave function, all of which give rise to the same measurement results. Moreover, a protection procedure frequently protects many different wave functions, but the protective measurement finds the one which is present.

2.2 Why I think that the quantum wave function describes a single quantum system (and everything else)

I want to believe that Science is capable of explaining everything. We should not explain every detail since we cannot store all the required information. I think it is enough to have a theory which can, given an unlimited storage and computational power, explain everything. The theory should agree with experiment. More specifically, any experiment, simple enough to allow us to predict its outcome theoretically, should be in agreement with the theory. Classical physics is not such a theory since many experiments like particle interference, atom stability, etc. contradict classical physics. Quantum mechanics is such a theory. There is no experiment today contradicting its prediction. The ontology of quantum theory is $|\Psi\rangle$ and this is why I believe it is real.

Accepting that there is no ontology beyond the wave function, we admit that every time we perform a preparation procedure of a particular quantum state, we end up with exactly the same situation. The failure of Bell inequalities, the PBR theorem and related results [8, 9] support this view. Then, a measurement on an ensemble of identically prepared systems can be viewed as a measurement of the wave function of each system. So, even without protective measurements one can accept the reality of the wave function of a single quantum system.

What might prevent us from considering quantum wave function as real is the collapse of the wave function at quantum measurement. Today there is no satisfactory physical explanation of the collapse and it is not plausible that such an explanation will be achieved due to non-locality and randomness of the collapse.

The approach according to which the wave function is not something real, but represents a subjective information, explains the collapse at quantum measurement perfectly: it is just a process of updating the information the observer has. This approach seems to be highly attractive. The problem is that it has not been successful until now. No one provided an answer to the question: "Information about what?" No good candidate for the underlying ontology has been proposed. Bell inequalities related to quantum measurements performed on entangled particles suggest that such an ontology, if we insist on locality, does not exist. Bell himself, however, never gave up looking for it. He even introduced the concept: local beables [10]. But in spite of much effort, no attractive theory of beables has been constructed.

One may say that a successful theory of *non-local* beables is provided by the de Broglie–Bohm theory; however, it does not make the wave function epistemic. Bohmians frequently say that "positions" represent primary ontology, but the wave function is still an ontology. And in some setups (surrealistic trajectories [11], weak measurements [12], protective measurements [13]) it is the wave function which provides an explanation of the observed results.

Quantum theory not only corresponds extremely well to our observations, it is also a very elegant theory except for the collapse of the wave function. In 1998 I heard Gottfried saying at a conference in Erice [14] that: "The reduction postulate is an ugly scar on what would be a beautiful theory if it could be removed." I firmly believe it can be removed. There is no experimental evidence for the collapse, only our prejudice that there are no multiple copies of every one of us. Removal of the collapse leads to the MWI. According to the MWI everything is a wave. The Universe is a highly entangled wave. It has natural decomposition into branches corresponding to different worlds in which macroscopic objects are described by well-localized wave functions. "Inside" a branch every photon emitted from a single photon source is described by its wave function; the same wave function which everyone will associate with photons emitted by a laser replacing the single photon source.

What are the alternatives? Only a minority remained with a hope of completing the ontology of quantum theory with "hidden variables". A consistent option is to accept that quantum theory is not about what Nature is, but about what we can say about it, as Bohr preached. This approach developed into a popular trend of Qbism [15], a kind of metaphysical nihilism: "In the quantum world, maximal information is not complete and cannot be completed." This nihilism seems unnecessary, because we are given an unprecedentedly successful deterministic theory of everything, the theory of a quantum wave function.

2.3 What is and what is not measurable using protective measurement

If we are given a single system in an unknown quantum state, there is no way to find out what this state is. It would contradict the no-cloning theorem: if we can find out what the state is, we can prepare many other systems in this state. However, if we are given a single system in a "protected" quantum state, we can find out its state.

In general, a single system might not have a description as a pure state. Formally, we can consider a situation when our system is entangled with another system, and the pure state of the composite systems is protected. Then we can observe the density matrix of our system which provides its complete description. However, if the systems are separate, then the locality of interactions disallows an efficient protection.

We can specify the wave function of the system by the expectation values of a set of variables. Thus, measuring these expectation values is equivalent to the measurement of the wave function. However, to argue that protective measurements help us in viewing the wave function of a single system as a "real" entity, we need weak measurements of projections on small regions of space to provide the spatial picture of the wave function directly. Also, weak measurements of local currents [16] allow specification of the phase of the wave function. Of course, only the relative phase can be found; the overall phase is unobservable.

If the wave function has a support in separate regions, the situation is less clear. Local currents do not allow us to reconstruct the relative phase between these regions. Another problem is the protection of such a state. Clearly, a local classical potential cannot lead to such protection: energy cannot depend on the relative phase. Although some non-local states can be measured using local interactions with the help of measuring devices with entangled parts [17], many other states cannot be measured in a non-demolition way [18].

A non-local state of equal superposition of spatially separated wave packets $\frac{1}{\sqrt{2}}(|A\rangle + |B\rangle)$ can be protected using non-local measurements. If it is a photon state it can be "swapped" to a Bell state of two spins [19]. Bell states can be measured in a non-demolition way using measurements of modular sums of spins $(\sigma_z^A + \sigma_z^B) \bmod 2$, $(\sigma_x^A + \sigma_x^B) \bmod 2$ [17]. Such measurements require measuring devices with entangled parts. Measurement of a fermion wave function requires also an antiparticle with known phase [20]. Even if the protection of a particular wave function is possible, it might not be enough for performing a protective measurement of this wave function. Non-local weak measurements are difficult and current proposals [21, 22] are not efficient. The superposition of wave packets of unequal weight, $\alpha|A\rangle + \beta|B\rangle$, $0 < |\alpha| < |\beta|$, cannot be measured in a non-demolition way [18], so in this case there is no protection procedure.

The main motivation for protective measurements was a slogan: "what is observable is real". I, however, do not fully support it. I do look for an ontology that can explain all of what we observe, but I do not want to constrain "observation" to instantaneous non-demolition measurements. I am ready to view the relative phase between two spatially separated parts of the wave function as "real", even if it can be observed only later, when the wave packets overlap.

2.4 The methods of protective measurements and the information gain

There are two methods for the protection of states: first, when it is a non-degenerate eigenstate of some Hamiltonian; second, based on the quantum Zeno effect, that frequent measurements of a variable for which the state is a non-degenerate eigenstate are performed. The strength of the protection is characterized by the energy gap to a nearby eigenstate and the frequency of measurements in the Zeno type protection. If the strength of the protection is known, the weakness parameter of weak measurements of the wave function can be calculated. Then, the whole wave function can be found. If the protection strength is not known, we can choose a weakness parameter, make the weak measurement of the wave function and then repeat it. If the result is the same, we know that the measurement is successful. If not, we should ask for another sample. Even if we need several systems, still, it is not a measurement on a large ensemble.

In the process of protective measurement we gain some information: the wave function is specified by many more parameters than the strength of the protection. There is also some gain of information beyond the information the party which arranges the protection must have. For protection it is enough to know that the state is one of the set of orthogonal states. After the protection and the weak measurement we will know which state of the set is given.

Probably, the Zeno-protection method is easier to understand. First, consider a measurement of the wave function on an ensemble. To know the absolute value of the wave function in a particular point, the projection on a small region of space around this point is measured, and the value is the ratio of the number of times the particle was found there to the number of trials. On every measurement, one of the eigenvalues of the projection operator, 0 or 1, is obtained. The statistical average provides the absolute value of the wave function there.

The projection measurement can be modeled by a von Neumann procedure in which at each trial a pointer, well localized at zero, shifts to 1 or remains at 0. When the uncertainty of the pointer is small, $\Delta \ll 1$, we know with certainty the outcome of each trial. In the corresponding weak measurement, the pointer remains at 0 if the particle is not there, and it is shifted by a small value ϵ, $\epsilon \ll \Delta$, if it is present there. The analysis of weak measurements shows that, up to a very good approximation, after the interaction with one system, the meter is shifted by $\epsilon|\psi|^2$. If we test now that the system remained in its original state, the probability of the failure is of the order of ϵ^2/Δ^2. Now, we repeat this procedure with an ensemble of $N = 1/\epsilon$ systems using the same measuring device. At the end, the pointer will show the desired value $|\psi|^2$ with almost unchanged uncertainty Δ. The probability that all the systems will be found in their original state after the interaction is $(1 - \epsilon^2/\Delta^2)^{1/\epsilon}$. If we make $\epsilon \ll \Delta^2$, the probability of even one failure goes to zero. It means that we can use the same system every time. This is the Zeno-type protective measurement.

Instead of the test by the projection on the state of interest, we can make measurements for which the state we want to measure is one of the eigenstates. If we know that we start with this state, we only need to perform measurement interactions without actually looking at the result. The procedure ensures that we will have our state for the duration of the procedure.

This procedure is different from a measurement on an ensemble not only because we use just one system. Although we have multiple couplings we do not get multiple outcomes and we do not calculate a statistical average as in the standard ensemble measurements.

The Zeno-type protection is more of a theoretical construction. Performing frequent verification measurements might be a very difficult if not impossible task. Another type of protection is to have a Hamiltonian with non-degenerate eigenstates. Experimentally, such a protection is much simpler. We have a system with a non-degenerate ground state. Then we just have to wait a long enough time (which can be reduced using some cooling procedure) and the wave function of the ground state is there and protected. It does not require much prior knowledge. Weak measurement coupling after a long enough time will provide the information about the wave function. Adiabatic switching of the weak coupling on and off will allow us to make the procedure faster.

In recent years, there has been great progress with cooling ions and atoms in traps. Science magazines frequently show pictures of what looks like a wave function of a trapped particle. But, as far as I understand, these pictures look like a "wave cloud" because of the width of the photons which are scattered by the ions. As far as I know, no protective measurements have been performed yet. Recent "wave function microscopy" [23] which used a photoionization imaging was also an ensemble measurement. Although I have no doubts about what the outcomes of protective measurements will be, I think it is of interest to make an effort to perform them, especially since many are reluctant to associate a wave function with a single system. I hope that some protective measurements will be performed in the near future, maybe along the lines of Nussinov's proposal [24].

The discussion above is about measurement of a stationary wave function. Protective measurements require some period of time to observe the wave function which is supposed to be constant during this period. Can it shed light on the question of the reality of a non-stationary wave function? If evolution is slow, unitary, and the protection is strong, then we can observe the wave function (with some limited precision) "in real time". If we start with the preselected state at the beginning of the protection procedure, it will ensure that the wave function will change in time appropriately and the weak coupling of the measurement of the wave will not disturb it significantly. A non-unitary evolution which includes "collapses" of the wave function cannot be observed in this way. The weak coupling will not be a problem, but we cannot ensure by any protection procedure that the wave function will collapse to the right state. This is one more reason not to believe that there is a collapse of a quantum state in Nature.

2.5 Protective measurement and postselection

Protective measurements support the approach according to which the reality of a quantum system is its wave function. Aharonov and I argued in many publications that the complete description of a quantum system is a two-state vector consisting of forward and backward evolving wave functions [25]. How can these apparently contradicting approaches peacefully coexist?

In the Zeno-type protective measurement, the backward evolving wave function is identical to the forward evolving wave function. Even if we postselect some other state, the last verification measurement "collapses" the backward evolving state to be identical to the forward evolving one. It cannot be any other eigenstate of the protection measurement since the measurement specifies both forward and backward evolving states.

For the Hamiltonian-type protection the situation is different. We may start, say, with a ground state of a harmonic oscillator and after performing a protective measurement of the wave function, measure the particle position. Whatever outcome we get, the backward evolving wave function will be very different from the forward evolving wave function. It will not be a stationary state, but an oscillating wave function.

Protective measurements cause almost no entanglement between the system and the measuring device, so the postselection according to the position of the particle should not change the outcomes of our protective measurements which constitute a picture of the forward evolving wave function. Weak coupling during a long period of time of the projection on every place x shows, at the end, the value of $|\psi(x)|^2$. For all outcomes of the postselection measurement we get the same (approximately correct) result of the density of the wave there. (The postselection which is very far from the center will correspond to some distorted picture since the weak coupling of protective measurements creates some small entanglement.) However, if we test whether during all the interaction time the coupling was to the same value, we will discover that this is not so (see Chapter 3). To make this test we should perform these protective measurements on a large pre- and postselected ensemble with all systems originally in the ground state. We should consider a sub-ensemble of particles postselected at localized state at position x. In this experiment, we will look at the pointer not at the end, but at various intermediate times. This will allow us to find the position of the pointer as a function of time. We will see that the pointer does not move in the same way during the time of the measurement. Sometimes it moves fast and sometimes it almost stops. It depends on the backward evolving wave function $\phi(x)$, which changes with time. The pointer "feels" the weak value of the projection operator $(\mathbf{P}_x)_w = \phi^*(x)\psi(x)/\langle\phi|\psi\rangle$ and it changes with time due to the time dependence of the backward evolving wave function.

This analysis suggests that the Hamiltonian protective measurement does not really measure the density of the wave function, but it measures a time average of a particle density which actually changes with time. To avoid this difficulty we can add a postselection of the state $|\psi\rangle$ at the end of the protective measurement. We know that it will succeed with certainty and then the weak measurement pointer will move with the same velocity during the whole process. Indeed, with the postselection, the weak measurements show a weak value of the pre- and postselected system. When the backward wave function is identical to the forward wave function, the weak value is equal to the expectation value, and it shows the density of the forward evolving wave function.

In the framework of the MWI, there is, however, a satisfactory explanation of the situation even without postselection of the original state $|\psi\rangle$. Let us consider again the case of the postselection measurement of position. Indeed, in every one out of the different worlds with postselection in different positions, the protective measurement measures the time average of the particle density as it is given by the weak value of the projection on a particular location. However, the time averages are subjective to the observers in worlds with different postselection positions. These are not objective realities. The objective reality is in the universe which incorporates worlds with all possible outcomes of the postselection measurement. In this world, the backward evolving wave function is identical to the forward evolving wave function [26]. To test this we can just look at the preselected ensemble only, without postselection. In this case the pointer will move with constant small velocity ending in the final value $|\psi(x)|^2$.

In the world with a postselection, subjective reality is best described by the two-state vector. The protective measurement described above showed just the forward evolving wave; we observe $|\psi(x)|^2$ and can also observe the local current defined by $\psi(x)$. This is because only the preselected forward evolving wave was protected. A different postselected state cannot be orthogonal (such postselection is impossible) and thus it cannot be protected by the same Hamiltonian or by the Zeno measurements. The backward evolving state $\phi(x)$ changes with time. The effective coupling to the pointer is through a weak value which oscillates, but after averaging, it roughly reproduces the expectation value corresponding to the forward evolving wave. If, instead of the protection Hamiltonian of the preselected state we arrange protection by another Hamiltonian (or frequent measurements) which protects the postselected backward evolving state, then our weak coupling of the measurement procedure will yield the backward evolving state $\phi(x)$. The forward evolving state will change in time and will lead to oscillating weak values averaging to the expectation values corresponding to $\phi(x)$.

At first, it seems that it is impossible to observe the two-state vector on a single system because the same Hamiltonian cannot protect both the forward evolving and the non-orthogonal backward evolving wave functions. Usually, all Hamitonians are Hermitian so they protect the same set of forward and backward evolving wave functions. However, if we pre- and postselect the whole system including the protection device, then for weak coupling the effective Hamiltonian will be the weak value of the Hamitonian. The weak value of the Hermitian operator might be a complex number. Then, the effective Hamiltonians responsible for evolution of the forward and the backward evolving states are different. Thus, we can, in this way, arrange protection of both forward and backward evolving wave functions even if they are different. This is the protection of the two-state vector [27].

The possibility of protecting the two-state vector does not necessarily mean that we can directly observe the two-state vector. The weak coupling to the projection on a particular location x will yield its weak value $(\mathbf{P}_x)_w = \phi^*(x)\psi(x)/\langle\phi|\psi\rangle$. Since weak measurements do not disturb each other significantly, we can measure in parallel weak values of other variables. Protection of the two-state vector will allow us then to view the full "weak measurement reality" [28] specified by the two-state vector. This will allow to reconstruct the two-state vector, but not in a direct way as it was in the protective measurement of the forward (or backward) evolving wave function.

I have to note that the protective measurement of the two-state vector reality is a highly theoretical concept. Its implementation in the laboratory requires a highly improbable result of the postselection measurement.

2.6 Critique of protective measurement

There have been many papers criticizing our work on protective measurement. According to Paraoanu [29] it has been established that the protective measurement program does not work: "Clearly, to think about the wave function as real, we would have to be able to measure it on a single quantum system. The question of whether this is possible was first raised in the context of the so-called "protective" (weakly disturbing) measurements in the early 1990s, where it was answered in the negative [2]." He relies on Alter and Yamamoto's paper [30] which analyzes the case when the quantum wave function is not protected and, not surprisingly, shows that in this case the wave function cannot be found [31]. Another reference of Paraoanu is "Impossibility of measuring the wave function of a single quantum system" by D'Ariano and Yuen [32]. Their claim is that an unknown and unprotected quantum state cannot be found (due to the no-cloning theorem). Regarding protective measurement they actually say that it works, given that the protection Hamiltonian is known. They add that the known protection Hamiltonian means that its eigenstates are known and that we can find which of the eigenstates is given without breaking the no-cloning theorem and without the need for a weak coupling of protective measurements. They did not distinguish between the party which can find the state and the party that protects the state and do not consider the case that all that is known is the strength of the protection of the state. Another work Paraoanu cites is a paper with a provocative title by Uffink: "How to protect the interpretation of the wave function against protective measurements."[33]. However, Gao recently criticized this paper [34] and Uffink retracted the main part of his objection [35].

There are also authors who praised protective measurements. Unruh [36] viewed protective measurement as a demonstration of the reality of certain operators and not of the wave function, but he admitted that "protective measurement has broadened our understanding of the quantum measurement process." Ghose and Home [37] in "An analysis of the Aharonov–Anandan–Vaidman model" wrote: "the AAV scheme serves to counteract the orthodox belief that quantum mechanics does not say anything empirically meaningful about an individual system." Dickson [38] considered protective measurement as "a good reply for the realist" against the empiricist.

It seems that many of the criticisms follow from misunderstanding triggered by the somewhat misleading examples in our PRA paper [4]. One of the examples was a particle in a superposition of two wave packets which is difficult to observe using protective measurement. The example was technically correct, since there was some tunneling wave connecting the wave packets such that they were only "almost" separate. The Stern–Gerlach experiment had its own difficulties, in particular, related to the divergenceless nature of the magnetic field.

Probably the most harsh criticism was a comment by Rovelli [39]: "We argue that the experiment does not provide a way for measuring noncommuting observables without a collapse, does not bear on the issue of the "reality of the wave function," and does not add any particular insight into our understanding (or non-understanding) of quantum mechanics." I hope, however, that after reading my paper the following quotation of Rovelli's comment makes his misunderstanding evident: "To make the problem particularly evident, consider the following experimental arrangement, which is entirely equivalent to the Aharonov–Anandan–Vaidman experiment as far as the interpretation of quantum mechanics is concerned. First, measure the polarization of a quantum particle. Second, write the outcome of this measurement on a piece of paper. At this stage assume that we do not know the polarization and we do not know what is written on the piece of paper. Then make the following protected measurement: Read what is written on the piece of paper."

After our clarification [40], Dass and Qureshi in a paper [41] "Critique of protective measurements" "looked at earlier criticisms of the idea, and concluded that most of them are not relevant to the original proposal." They argued, however, that "one can never perform a protective measurement on a single quantum system with absolute certainty. This clearly precludes an ontological status for the wave function." I agree that protective measurements do not provide absolute certainty, but as I explained above, this does not prevent me from attributing an ontological status to the wave function.

Protective measurements do not provide a decisive argument for the ontology of the wave function. However, they definitely provide a deep insight into the process of quantum measurement and they strengthen significantly the realist interpretation of the wave function of a single particle. I hope that some protective measurements will be performed in the near future. I believe that they will lead to a significant progress in understanding quantum reality.

This work has been supported in part by grant number 32/08 of the Binational Science Foundation and the Israel Science Foundation Grant No. 1125/10.

References

[1] Y. Aharonov and L. Vaidman, Properties of a quantum system during the time interval between two measurements, *Phys. Rev. A* **41**, 11 (1990).

[2] L. Vaidman, Many-worlds interpretation of quantum mechanics, *Stan. Enc. Phil.*, E. N. Zalta (ed.) (2002), plato.stanford.edu/entries/qm-manyworlds/.

[3] Y. Aharonov and L. Vaidman, The Schrödinger wave is observable after all! In *Quantum Control and Measurement*, H. Ezawa and Y. Murayama (eds.), 99 (Tokyo: Elsevier Publishing, 1993).

[4] Y. Aharonov, J. Anandan, and L. Vaidman, Meaning of the wave function, *Phys. Rev. A* **47**, 4616 (1993).

[5] Y. Aharonov and L. Vaidman, Measurement of the Schrödinger wave of a single particle, *Phys. Lett. A* **178**, 38 (1993).

[6] M. F. Pusey, J. Barrett, and T. Rudolph, On the reality of the quantum state, *Nature Phys.* **8**, 476 (2012).

[7] J. S. Lundeen, B. Sutherland, A. Patel, C. Stewart, and C. Bamber, Direct measurement of the quantum wavefunction, *Nature* **474**, 188 (2011).

[8] R. Colbeck and R. Renner, Is a system's wave function in one-to-one correspondence with its elements of reality? *Phys. Rev. Lett.* **108**, 150402 (2012).

[9] L. Hardy, Are quantum states real? *Int. J. Mod. Phys. B*, **27**, 1345012 (2013).

[10] J. S. Bell, Beables for quantum field theory, in J. S. Bell, *Speakable and Unspeakable in Quantum Mechanics,* pp. 173–180 (Cambridge: Cambridge University Press, 1987).

[11] B. G. Englert, M. O. Scully, G. Süssmann, and H. Walther, Surrealistic Bohm trajectories, *Z. Naturforsch. A* **47**, 1175 (1992).

[12] Y. Aharonov and L. Vaidman, About position measurements which do not show the Bohmian particle position, in *Bohmian Mechanics and Quantum Theory: an Appraisal*, J. T. Cushing, A. Fine and S. Goldstein (eds.), pp. 141–154 (Dordrecht: Kluwer, 1996).

[13] Y. Aharonov, M. O. Scully, and B. G. Englert, Protective measurements and Bohm trajectories, *Phys. Lett. A* **263**, 137 (1999).

[14] K. Gottfried, Does quantum mechanics describe the "collapse" of the wave function?, contribution to the Erice school: *62 Years of Uncertainty* (unpublished) (1989).

[15] C. M. Caves, C. A. Fuchs, and R. Schack, Quantum probabilities as Bayesian probabilities, *Phys. Rev. A* **65**, 022305 (2002).

[16] M. V. Berry, Five momenta, *Eur. J. Phys.* **34**, 1337 (2013).

[17] Y. Aharonov, D. Z. Albert, and L. Vaidman, Measurement process in relativistic quantum theory, *Phys. Rev. D* **34**, 1805 (1986).

[18] S. Popescu and L. Vaidman, Causality constraints on nonlocal quantum measurements, *Phys. Rev. A* **49**, 4331 (1994).

[19] L. Vaidman, Nonlocality of a single photon revisited again, *Phys. Rev. Lett.* **75**, 2063 (1995).

[20] Y. Aharonov and L. Vaidman, Nonlocal aspects of a quantum wave, *Phys. Rev. A* **61**, 052108 (2000).

[21] K. J. Resch and A. M. Steinberg, Extracting joint weak values with local, single-particle measurements, *Phys. Rev. Lett.* **92**, 130402 (2004).

[22] A. Broduch and L. Vaidman, Measurements of non local weak values, *J. Phys.: Conf. Ser.* **174**, 012004 (2009).

[23] S. Cohen, M. M. Harb, A. Ollagnier, et al., Wave function microscopy of quasibound atomic states, *Phys. Rev. Lett.* **110**, 183001 (2013).

[24] S. Nussinov, Scattering experiments for measuring the wave function of a single system, *Phys. Lett. B* **413**, 382 (1997).

[25] Y. Aharonov and L. Vaidman, Complete description of a quantum system at a given time, *J. Phys. A: Math. Gen.* **24**, 2315 (1991).

[26] L. Vaidman, Time symmetry and the many-worlds interpretation, in *Many Worlds? Everett, Quantum Theory, and Reality*, S. Saunders, J. Barrett, A. Kent, and D. Wallace (eds.) (Oxford: Oxford University Press, 2010).

[27] Y. Aharonov, and L. Vaidman, Protective measurements of two-state vectors, in *Potentiality, Entanglement and Passion-at-a-Distance*, R. S. Cohen, M. Horne and J. Stachel (eds.), BSPS 1–8, (Dordrecht: Kluwer, 1997).

[28] L. Vaidman, Weak-measurement elements of reality, *Found. Phys.* **26**, 895 (1996).

[29] G. S. Paraoanu, Extraction of information from a single quantum, *Phys. Rev. A* **83**, 044101 (2011).

[30] O. Alter and Y. Yamamoto, Protective measurement of the wave function of a single squeezed harmonic-oscillator state, *Phys. Rev. A* **53**, R2911 (1996); *Phys, Rev. A* **56**, 1057 (1997).

[31] Y. Aharonov and L. Vaidman, Protective measurement of the wave function of a single squeezed harmonic-oscillator state – comment, *Phys. Rev. A* **56**, 1055 (1997).

[32] G. M. D'Ariano and H. P. Yuen, Impossibility of measuring the wave function of a single quantum system, *Phys. Rev. Lett.* **76**, 2832 (1996).

[33] J. Uffink, How to protect the interpretation of the wave function against protective measurements, *Phys. Rev. A* **60**, 3474 (1999).

[34] S. Gao, On Uffink's criticism of protective measurements, *Stud. Hist. Phil. Mod. Phys.* **44**, 513 (2013).

[35] J. Uffink, Reply to Gao's "On Uffink's criticism of protective measurements" *Stud. Hist. Phil. Mod. Phys.* **44**, 519 (2013).

[36] W. G. Unruh, Reality and measurement of the wave function, *Phys. Rev. A* **50**, 882 (1994).

[37] P. Ghose and D. Home, An analysis of the Aharonov–Anandan–Vaidman model, *Found. Phys.* **25**, 1105 (1995).

[38] M. Dickson, An empirical reply to empiricism: protective measurement opens the door for quantum realism, *Phil. Sci.* **62**, 122 (1995).

[39] C. Rovelli, Meaning of the wave function – comment, *Phys. Rev. A* **50**, 2788 (1994).

[40] Y. Aharonov, J. Anandan, and L. Vaidman, The meaning of protective measurements, *Found. Phys.* **26**, 117 (1996).

[41] N. D. H. Dass and T. Qureshi, Critique of protective measurements, *Phys. Rev. A* **59**, 2590 (1999).

3

Protective measurement, postselection and the Heisenberg representation

YAKIR AHARONOV AND ELIAHU COHEN

Classical ergodicity retains its meaning in the quantum realm when the employed measurement is protective. This unique measuring technique is re-examined in the case of post-selection, giving rise to novel insights studied in the Heisenberg representation. Quantum statistical mechanics is then briefly described in terms of two-state density operators.

3.1 Introduction

In classical statistical mechanics, the ergodic hypothesis allows us to measure position probabilities in two equivalent ways: we can either measure the appropriate particle density in the region of interest or track a single particle over a long time and calculate the proportion of time it spent there. As will be shown below, certain quantum systems also obey the ergodic hypothesis when protectively measured. Yet, since Schrödinger's wave function seems static in this case [1, 2, 3], and Bohmian trajectories were proven inappropriate for calculating time averages of the particle's position [4, 5], we will perform our analysis in the Heisenberg representation.

Indeed, quantum theory has developed along two parallel routes, namely the Schrödinger and Heisenberg representations, later shown to be equivalent. The Schrödinger representation, due to its mathematical simplicity, has become more common. Yet, the Heisenberg representation offers some important insights which emerge in a more natural way, especially when employing modular variables [6]. For example, in the context of the two-slit experiment it sheds a new light on the question of momentum exchange [7, 8, 9]. Recently studied within the Heisenberg representation are also the double Mach–Zehnder interferometer [10] and the N-slit problem [11]. As can be concluded from [11], the Heisenberg representation prevails in emphasizing the non-locality in quantum mechanics, thus providing us with insights about this aspect of quantum mechanics as well.

Equipped with the backward evolving state-vector within the framework of two-state-vector formalism (TSVF) [12], the Heisenberg representation becomes even more powerful since the time evolution of the operators includes now information from the two boundary conditions. Furthermore, when performing post-selection, deeper understanding of the quantum system becomes available, such as the past of a quantum particle [13, 14].

Post-selection does not change the protective measurement's results, but suggests interpreting them differently, thus enabling us to effectively sketch two wave functions rather than one in the Schrödinger representation. In the Heisenberg representation, a full description of time-dependent operators emerges which enables further insights. Choosing a specific final state amounts to outlining another (sometimes, completely different) history for the same initial state, that is, a different set of characterizing weak values. In what follows, we use the Heisenberg representation to study protective measurements with post-selection. This way, we regain quantum ergodicity and describe two-state ensembles coupled to a heat bath.

The rest of the chapter is organized as follows: Section 3.2 discusses the differences between classical and quantum ergodicity. Section 3.3 describes protective measurement in the Heisenberg representation. Cases of post-selection and external protection are analyzed. In Section 3.4 we show how to describe quantum statistical mechanics in terms of two-state vectors. Protective measurement is utilized for studying the two-state density operator and the resulting ensemble averages. Section 3.5 summarizes the main contributions of this work into a coherent description of protective measurement in the Heisenberg representation.

3.2 Classical and quantum ergodicity

We begin by examining a classical gas, i.e. an ensemble of N point-like particles. Each individual particle is characterized by its position and momentum, so that in each moment the system can be described by a point in $6N$-dimensional phase-space. The time average of a certain property A over a time interval of length T is given by:

$$\bar{A} = \lim_{n \to \infty} \frac{1}{n} \sum_{j=1}^{n} A\left(\frac{jT}{n}\right). \tag{3.1}$$

Therefore, in order to accurately find $\bar{A}$ we ought to perform a large number of A measurements at different times.

Under the ergodic assumption [15] this average is equivalent to the ensemble average at a certain moment:

$$\langle A \rangle = \sum_{j=1}^{N} \frac{A_j}{N}.$$ (3.2)

More generally,

$$\langle A \rangle = \int A d\mu,$$ (3.3)

where μ is some finite, non-zero probability measure.

Is this reasoning applicable also in the quantum realm? First, in order to incorporate uncertainty, the phase-space should be partitioned into hypercubes of volume $\hbar^{6N}$. Second, a practical question has to be addressed: how to perform all the measurements needed for an accurate time average on a single particle without disturbing it? This is where a resolution can be achieved with the help of protective measurement suggested for the first time by Aharonov and Vaidman in 1993 [1] and further developed in [2, 3, 16, 17]. Moreover, using protective measurement it was argued that the wave function should be understood as describing the (discontinuous, random in nature) ergodic motion of a single particle [18].

3.3 Protective measurement in the Schrödinger and Heisenberg representations

Protection of the state in the case of discrete non-degenerate spectrum of energy eigenstates was shown to be a consequence of energy conservation when the measurement is sufficiently slow and weak [2]. Protection can be achieved also in more general cases by utilizing a protective interaction term in the Hamiltonian. This possibility of performing a dense set of measurements without affecting the measured state, allowed "observing" of the wave function [1]. In the Schrödinger representation it seems that the evolution of the wave function was tightly restricted, which let us later obtain its form everywhere in space. Putting it in more formal terms, protective measurement can be carried out by applying an interaction Hamiltonian of the form:

$$H_{\text{int}} = g(t)pP_{V_i},$$ (3.4)

with $g(t) = 1/T$ for a period of T smoothly approaching zero before and after the measurement, where p is the momentum of the measuring pointer, P_{V_i} is the projection operator into the set V_i, and $V = \sum_i V_i$ is the total space region. Let us assume that the system in question is a harmonic oscillator, and the initial wave function is the ground state $|\psi_{in}\rangle = |0\rangle$, i.e. $\psi(x) = \pi^{-1/4}e^{-x^2/2}$ (throughout the calculations we used $\sqrt{\hbar/m\omega} = 1$).

Suppose also that we are interested in some remote V_i centered around $x_0 \gg 1$, i.e. far from the origin. The particle has a small probability of being found in that place, but when the measurement is long enough, we would find that the state of the pointer propagated in time according to:

$$U = e^{-\frac{i}{\hbar}p\langle P_{V_i}\rangle}, \tag{3.5}$$

although the energy has only changed negligibly for each p:

$$\delta E = \langle H_{int}\rangle = \frac{\langle P_{V_i}\rangle p}{T}. \tag{3.6}$$

This way we can gain knowledge of $|\psi_{in}\rangle$ of a single particle in V_i. Repeating this measurement for all V_i we would finally be able to sketch $|\psi_{in}\rangle$ in V.

Here we introduce post-selection in the form of slicing past events using a certain final state [19, 20]. By this we mean grouping together all the experiments which ended at the same state. What does it change? Clearly, the results of the protective measurement do not change, giving rise to the same observation of the wave function. The ontology, however, turns out to be different. Our initial state was $|\psi_{in}\rangle = |0\rangle$. When performing the trivial post-selection, that is, $|\psi_{fin}\rangle = |\psi_{in}\rangle$, within the Schrödinger representation we believe that the protective measurement probed a static (up to a changing phase) eigenstate of the oscillator having a small probability of being found in V_i. Hence, the pointer translation grew slowly but surely according to Eq. (3.5). However, suppose we post-select a different final state which is some coherent state $|\alpha\rangle$ (since coherent states form an overcomplete basis, this can be done approximately by defining the appropriate POVM). In our experiment, the final measurement will allow finding of the initial state as a coherent state $|\alpha\rangle$ with probability $e^{-|\alpha|^2/2}$. In the position representation, the coherent state is denoted at every moment by [21]:

$$\varphi_\alpha(x, t) = \pi^{-1/4} \exp\{-i\Theta(x, t) - \frac{1}{2}[x - \sqrt{2}|\alpha|\cos(\omega t - \delta)]^2\}, \tag{3.7}$$

where $\alpha = |\alpha|e^{i\delta}$ and

$$\Theta(x, t) = \frac{\omega t}{2} - \frac{|\alpha|^2}{2}\sin[2(\omega t - \delta)] + \sqrt{2}|\alpha|x\sin(\omega t - \delta). \tag{3.8}$$

The same result of Eq. (3.5) suggests now a significant motion along the harmonic well of this backward evolving coherent state. As was shown in [22, 23], any sufficiently weak coupling between a pointer and an observable O of a pre- and post-selected quantum system is a coupling to a weak value:

$$O^w(t) = \frac{\langle \Phi_f(t)|O|\Phi_i(t)\rangle}{\langle \Phi_f|\Phi_i\rangle}, \qquad (3.9)$$

where $|\Phi_i\rangle$ and $\langle\Phi_f|$ are the pre- and post-selected states respectively. In order to demonstrate the movement of the pointer we shall assume its coupling to the real part of the weak value and find out:

$$\mathrm{Re}\{P_{V_i}^w(t)\} = \mathrm{Re}\left\{\frac{\langle\alpha^*(t)|P_{V_i}|\psi_{\mathrm{in}}(t)\rangle}{\langle\alpha^*|\psi_{\mathrm{in}}\rangle}\right\}, \qquad (3.10)$$

that is:

$$\mathrm{Re}\{P_{V_i}^w(t)\} \approx \pi^{-1/2}e^{(|\alpha|^2-x_0^2)/2}\cos[\Xi(t)]e^{-[x_0-\sqrt{2}|\alpha|\cos(\omega(T-t)+\delta)]^2/2}, \qquad (3.11)$$

where

$$\Xi(t) = \frac{\omega T}{2} + \frac{|\alpha|^2}{2}\sin[2(\omega(T-t)+\delta)] - \sqrt{2}|\alpha|x_0\sin[\omega(T-t)+\delta]. \qquad (3.12)$$

Due to the oscillations of the post-selected coherent state, the pointer translation can be understood now to be non-linear. According to Eq. (3.11) the pointer movement seems oscillatory (it moves each time the backward evolving coherent state "pushes" it), which is quite different from the case of trivial post-selection where it moved linearly, so it finally reaches the same place as earlier, but with an altered history. A comparison between the expectation value of the pointer readings in the case of trivial post-selection and in the case of α post-selection is shown in Fig. 3.1. For illustration purposes, the following parameters were chosen: $x_0 = 1$, $\alpha = 2.5$, $\omega = 1$ Hz and $T = 100$ s. We assume that the width of the pointer's wave function is large enough so that the measurement can be considered weak.

In order to better understand the movement which arises from Eq. (3.11) we compare the results of the above $\alpha = 2.5$ post-selection to post-selection of $\alpha = 1$ (while the other parameters remain the same). The forward and backward evolving states are now closer, so due to their higher scalar product, the weak value, and hence the amplitude of oscillations, both decrease (see Fig. 3.2). Another comparison is drawn between the above case of searching for the wave function at $x_0 = 1$ to the case of searching at $x_0 = 1.5$. The chances of finding the particle there are now smaller, and therefore, the expectation value is lower (see Fig. 3.3).

Utilizing Bohr's correspondence principle, we could relate classical and quantum ergodicity: if instead of the ground state we had chosen a highly excited state (or alternatively, large α for the final state), we know, according to the

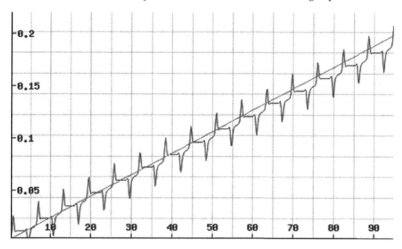

Figure 3.1 Pointer readings for two post-selections. The pointer readings are shown for the trivial post-selection (linear graph) and for the α post-selection (oscillatory graph). Despite the different shape, they eventually reach approximately the same point.

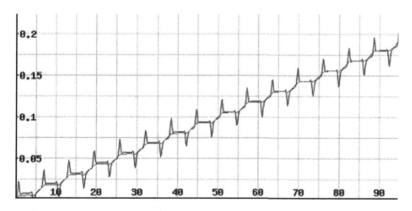

Figure 3.2 A comparison between $\alpha = 2.5$ (high amplitude) and $\alpha = 1$ (low amplitude).

correspondence principle, that the classical time the oscillator spends in V_i would be proportional to the relative number of harmonic oscillators, out of a large ensemble, that could be found instantaneously within this interval.

This dynamic interpretation can be better understood within the Heisenberg representation. First, we know that the operators $\hat{x}$ and $\hat{p}$ change in time just like the classical variables x and p, hence ergodicity and correspondence arise naturally.

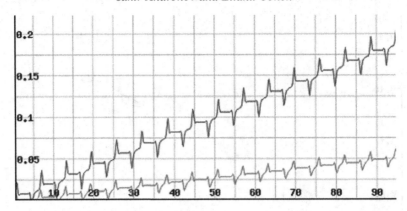

Figure 3.3 A comparison between $x_0 = 1$ (upper) and $x_0 = 1.5$ (lower).

Second, each projection operator $P_{V_i}(t)$ can be evaluated as a time-dependent matrix using the oscillator eigenstates:

$$P_{V_i}^{m,n}(t) = e^{-i(m-n)t}\langle m|P_{V_i}|n\rangle, \tag{3.13}$$

which in contrast to the evolution of the state seems very oscillatory. However, during the measurement interval, all the off-diagonal entries tend to zero, and P_{V_i} becomes approximately time-independent and diagonal. Therefore, after a long time its diagonal values directly indicate ensemble averaging, thus expressing quantum ergodicity. This could also be understood from the coherent states evolution which covers all phase space, thus allowing the operators in the Heisenberg representation to take any possible value. Slicing past results according to all the possible future results divides the ensemble to several distinct sub-ensembles, each of which has a different weak value and hence a different history of the measuring pointer.

Another discrepancy between the two representations apparently arises in the case when the initial state is a superposition of different energy eigenstates. Artificial Zeno-type protection is needed in the form of very frequent projective measurement on the state, which will preserve it by halting its evolution (the time scale of intervals between consecutive protections must be much smaller than the time scale of changing the wave function due to its Hamiltonian). In the Schrödinger representation, it seems that the state rarely changes due to this procedure; hence protective measurements are performed again and again on one and the same static state. In contrast, calculation in the Heisenberg representation describes the image of subsequent abrupt changes of the operator we wish to measure.

3.4 Statistical mechanics with two-state vectors

Assume now the system is coupled to a heat bath of temperature $T = (k\beta)^{-1}$ and allowed to reach equilibrium. The system will be described by the Boltzmann thermal density operator:

$$\rho = \frac{e^{-\beta H}}{\text{Tr}(e^{-\beta H})}. \tag{3.14}$$

For the harmonic oscillator discussed above it equals [24]:

$$\rho = (1 - e^{-\beta\hbar\omega}) \sum_{n=0}^{\infty} e^{-n\beta\hbar\omega}|n\rangle\langle n|. \tag{3.15}$$

If the measuring time is longer than the period of thermal fluctuations, the protective measurement will indicate the correct mixed state; that is, the pointer will move according to the thermal average of the measured quantity. Alternatively, one can switch off the coupling to the thermal bath before performing the measurement, and then the measurement will select a single pure state, rather than a mixture, according to the Boltzmann distribution.

Recalling the mapping between the averages calculated with this operator and the expectation values of the pure state [24]:

$$|\psi_\beta\rangle = 1 - e^{-\beta\hbar\omega^{1/2}} \sum_{n=0}^{\infty} e^{-n\beta\hbar\omega/2}|n\rangle, \tag{3.16}$$

we can perform protective measurements of this state and find out expectation values of thermal ensembles without disturbing them. A single protective measurement was shown until now to describe the wave function of a single particle, and here it allows us to acquire knowledge about a large ensemble coupled to a heat bath.

What is the time-symmetric version of this density operator? The TSVF [12] enables us to describe a quantum system in between two strong measurements with the aid of weak measurements [22]. It is a symmetric formulation of quantum mechanics ascribing equal footing to the initial (forward evolving) and final (backward evolving) wave functions. The two-state vector $\langle\Phi|\ |\Psi\rangle$ was shown in [25] to give rise to the density operator:

$$\rho(t) = \frac{|\Psi(t)\rangle\langle\Phi(t)|}{\langle\Phi|\Psi\rangle}, \tag{3.17}$$

which evolves according to the von Neumann equation just like the one-state density operator:

$$i\hbar\frac{\partial\rho}{\partial t} = [H, \rho]. \tag{3.18}$$

In the double coordinate system it was shown to be:

$$i\hbar\frac{\partial\rho(x',x'',t)}{\partial t} = [H(x',p') - H(x'',p'')]\rho(x',x'',t), \qquad (3.19)$$

where $\rho(x',x'',t) = \langle x'|\rho(t)|x''\rangle$.

The two-state density operator enables the calculation of weak values as follows:

$$A_{\mathrm{w}} = \frac{\mathrm{Tr}(A\rho)}{\mathrm{Tr}(\rho)}. \qquad (3.20)$$

Examining now a canonical ensemble with inverse temperature $T = (k\beta)^{-1}$, the two-state density ρ would take the form:

$$\rho = \frac{\exp\{-\beta[H(x',p') - H(x'',p'')]\}}{\mathrm{Tr}(\exp\{-\beta[H(x',p') - H(x'',p'')]\})}, \qquad (3.21)$$

thus allowing us to calculate ensemble- and hence time-averages in the two-state Heisenberg representation when employing protective measurement.

3.5 Discussion

The wave function as observed by protective measurement gains its meaning only when very long measurements or measurements over a large ensemble are performed. It is not possible to measure instantaneously the wave function of a single particle. This suggests that the wave function has either a statistic or an ergodic meaning. However, operators in the Heisenberg representation do allow a description of a single quantum particle at a single time. In addition, when pre- and postselection are performed, the measuring pointer describes a distinct history of the system, depending on both backward and forward evolving wave functions. Furthermore, a single protective measurement allows us to find the thermal state of an ensemble coupled to a heat bath, which leads to a full description of two-state thermal ensembles.

Acknowledgements

We wish to thank Avshalom C. Elitzur, Tomer Landsberger and Daniel Rohrlich for helpful comments and discussions. This work has been supported in part by the Israel Science Foundation Grant No. 1125/10.

References

[1] Y. Aharonov and L. Vaidman, The Schrödinger wave is observable after all! In *Quantum Control and Measurement*, H. Ezawa and Y. Murayama (eds.), Tokyo: Elsevier Publications (1993).

[2] Y. Aharonov, J. Anandan and L. Vaidman, Meaning of the wave function, *Phys. Rev. A* **47**, 4616 (1993).

[3] Y. Aharonov and L. Vaidman, Measurement of the Schrödinger wave of a single particle, *Phys. Lett. A* **178**, 38 (1993).

[4] Y. Aharonov, M. O. Scully and B. G. Englert, Protective measurements and Bohm trajectories, *Phys. Lett. A* **263**, 137 (1999).

[5] Y. Aharonov, N. Erez and M. O. Scully, Time and ensemble averages in Bohmian mechanics, *Physica Scripta* **69**, 81–83 (2004).

[6] Y. Aharonov, H. Pendleton and A. Petersen, Modular variables in quantum theory, *Int. J. Th. Phys.* **2**, 213 (1969).

[7] M. O. Scully, B. G. Englert and H. Walther, Quantum optical tests of complementarity, *Nature* **351**, 111–116 (1991).

[8] S. Durr, T. Nonn and G. Rempe, Origin of quantum-mechanical complementarity probed by a 'which-way' experiment in an atom interferometer, *Nature* **395**, 33–37 (1998).

[9] T. J. Herzog, P. G. Kwiat, H. Weinfurter and A. Zeilinger, Complementarity and the quantum eraser, *Phys. Rev. Lett.* **75**, 3034–3037 (1995).

[10] J. Tollaksen, Y. Aharonov, A. Casher, T. Kaufherr and S. Nussinov, Quantum interference experiments, modular variables and weak measurements, *New J. Phys.* **12**, 013023 (2010).

[11] Y. Aharonov, On the Aharonov–Bohm effect and why Heisenberg captures nonlocality better than Schrödinger. In *Memory of Akira Tonomura: Physicist and Electron Microscopist*, Y. A. Ono and K. Fujikawa (eds.), Singapore: World Scientific Publishing Company (2013).

[12] Y. Aharonov and L. Vaidman, The two-state vector formalism of quantum mechanics, in *Time in Quantum Mechanics*, J. G. Muga *et al.* (eds.), Berlin: Springer, 369–412 (2002).

[13] L. Vaidman, Past of a quantum particle, *Phys. Rev. A* **87**, 052104 (2013).

[14] A. Danan, D. Farfurnik, S. Bar-Ad and L. Vaidman, Asking photons where have they been, *Phys. Rev. Lett.* **111**, 240402 (2013).

[15] K. G. Kay, Toward a comprehensive semiclassical ergodic theory, *J. Chem. Phys.* **79**, 3026 (1983).

[16] Y. Aharonov and L. Vaidman, Protective measurements of two-state vectors. In *Potentiality, Entanglement and Passion-at-a-Distance*, R. S. Cohen, M. Horne and J. Stachel (eds.), BSPS 1–8 (Dordrecht: Kluwer, 1997).

[17] Y. Aharonov, J. Anandan and L. Vaidman, The meaning of protective measurements, *Found. Phys.* **26**, 117–126 (1996).

[18] S. Gao, Meaning of the wave function, *Int. J. Quantum Chem.* **111**, 4124–4138 (2011).

[19] Y. Aharonov, E. Cohen and A. C. Elitzur, Foundations and applications of weak quantum measurement, *Phys. Rev. A* **89**, 052105 (2014).

[20] Y. Aharonov, E. Cohen, D. Grossman and A. C. Elitzur, Can weak measurement lend empirical support to quantum retrocausality, *EPJ Web of Conferences* **58**, 01015 (2013).

[21] F. Schwabl, *Quantum Mechanics*, 3rd edition, Berlin: Springer, 54–56 (2002).

[22] Y. Aharonov, D. Albert and L. Vaidman, How the result of a measurement of a component of a spin 1/2 particle can turn out to be 100, *Phys. Rev. Lett.* **60**, 1351–1354 (1988).

[23] Y. Aharonov, E. Cohen and S. Ben-Moshe, Unusual interactions of pre- and post-selected particles, Proceedings of 1st International Conference on New Frontiers in Physics, *EPJ Web of Conferences* **70**, 00053 (2014).

[24] Y. Aharonov, E. C. Lerner, H. W. Huang and J. M. Knight, Oscillator phase states, thermal equilibrium and group representations, *J. Math. Physics* **14**, 746–756 (1973).

[25] B. Reznik and Y. Aharonov, Time-symmetric formulation of quantum mechanics, *Phys. Rev. A* **52**, 2538–2550 (1995).

4

Protective and state measurement: a review

GENNARO AULETTA

In this chapter, I summarize the general way in which the measurement process can be cast (by making use of effects and amplitude operators). Then, I show that there are two main problems with the state measurement: (i) how to avoid the disruptive back action on the state of the measured system during detection and (ii) how to extract complete information from this state. In order to deal with them I first introduce quantum non-demolition (QND) measurement and examine the problem whether the entire probability distribution of the measured observable is not altered by a QND measurement, which would allow repeated QND measurements with different observables to extract the whole information from the measured system. However, I show that this is not the case. Then, I deal with protective measurement as such and show that a reversible measurement is in fact not a measurement. However, by taking advantage of statistical methods (and therefore by renouncing measurement of the state of a single system), we can indeed reconstruct the wave function but only by partially recovering the information contained in the state. Two further prices to pay are to admit the existence of negative quasi-probabilities due to the interference terms and to make use of unsharp observables for guaranteeing informational completeness.

4.1 Introduction

The aim of this chapter is a review of the developments related to the measurement of the state vector and the problems that have been raised in this context.

Traditionally, the state vector or the wave function describing quantum systems has been considered as a formal tool for calculating the probabilities associated with certain events like measurement outcomes, but few scholars have tried to attribute to it an ontological status. It is well known how many difficulties are related to an attempt at assigning a kind of reality to the state of quantum systems.

39

Nevertheless, a major concern of some physicists has been precisely the query about the physical significance of this entity. If we want to approach this problem, it is quite natural to ask whether there is a way to measure the state vector. In fact, if we assume that this is kind of reality, we should be able to find certain experimental contexts in which it makes a difference to make such an ontological attribution [3, chapter 28].

In the following, I first present a general formalism for measurement (Section 4.2), then I introduce the quantum non-demolition measurement (Section 4.3). The core of the chapter is represented by an analysis of protective measurement (Section 4.4). On such a basis, the issue about the relation between measurement and reversibility is raised (Section 4.5). In Section 4.6 the method for reconstructing the wave function of a quantum system is shown, while in Section 4.7 the prices to pay are summarized. Finally, in Section 4.8 I draw some synthetic conclusions.

4.2 Measurement in general

We recall that quantum mechanically, in order to recover the information about an object system, we need the coupling with an apparatus. I shall show now that classically we have a similar situation. Here, we have an unknown parameter k whose value we wish to know and some data d pertaining to a set D at our disposal. This is a very important point, since we *never* have direct access to things (whose properties are described by k) but always to things *through* data [5, chapter 2]. These data can be represented by the position of the pointer of our measuring apparatus or simply by the impulse our sensory system has received, or even by the way we receive information about the position of the pointer through our sensory system. It does not matter how long this chain may be. The important point is a matter of principle: we can receive information about objects and events only conditionally on the data at our disposal. Let us consider a classical example. Suppose that we wished to know exactly what the distribution of matter was in the early Universe. We can know this by collecting data about the cosmic microwave background radiation we receive now. This again shows a very important common point between quantum and classical physics that is not well understood, and which has been pointed out by Wheeler's delayed choice thought-experiment [25]. We cannot receive any information about past events unless they are received through *present effects* (data). This is an equivalent formulation of what we have said before, since any event, represented by a parameter k, can be known only through a *later* effect due to the finite speed of light. As a matter of fact, all of our perceptual experience is mediated and slightly delayed in time. Moreover, we always have experience only of a part of the possible effects produced by events.

Obviously, once we have observed or acquired data, we must perform an information extrapolation that allows us to have an "informed guess" about the value k of the parameter. This is the process of information selection. As we know, the joint probability $\wp(j,k)$ that we select the event j while having an event represented by the value k of an unknown parameter (i.e., the probability that both event k and event j occur) is given by

$$\wp(j,k) = \wp(j|k)\wp(k), \qquad (4.1)$$

where $\wp(j|k)$ is the condition probability of the selection event j given the source event represented by k. Now, by taking into account the data d that are somehow the interface between the source event k and our selection event j we may express the probability $\wp(j|k)$ as [16]

$$\wp(j|k) = \sum_{d \in D} \wp(j|d)\wp(d|k), \qquad (4.2)$$

where the summation is over all the data d pertaining to the set D. By substituting the above expression into Eq. (4.1) we obtain

$$\begin{aligned}
\wp(j,k) &= \sum_{d \in D} \wp(j|d)\wp(d|k)\wp(k) \\
&= \sum_{d \in D} \wp(j|d)\wp(d,k). \qquad (4.3)
\end{aligned}$$

We note that Eq. (4.3) can be considered as a generalization of the well-known formula

$$\wp(j) = \sum_{d \in D} \wp(j|d)\wp(d), \qquad (4.4)$$

and it reduces to the latter when $\wp(k) = 1$, i.e., when the value k of the parameter is known with certainty. It is important to stress that the two conditional probabilities $\wp(j|d)$ and $\wp(d|k)$ are quite different. The probability $\wp(d|k)$ represents how *faithful* our data are given the source event k, that is, how reliable our apparatus (or sensory system) is. Instead, the probability $\wp(j|d)$ represents our ability to select a *single* event j which can be used to interpret the data d in the best way. Moreover, using the Bayes theorem we express $\wp(k|j)$ in terms of $\wp(j|d)$ and $\wp(d|k)$ as

$$\begin{aligned}
\wp(k|j) &= \frac{\wp(k)\wp(j|k)}{\wp(j)} \\
&= \frac{\wp(k)}{\wp(j)} \sum_{d \in D} \wp(j|d)\wp(d|k). \qquad (4.5)
\end{aligned}$$

In other words, we can invert the kind of question we pose and try to infer the unknown value k of the parameter conditioned on having selected the event j.

Having made these considerations, we immediately see that Eq. (4.2) or (4.3) represents the classical analogue of the quantum measurement process. The conditional probability $\wp(d|k)$ corresponds to the coupling between the object system and the apparatus in quantum mechanics. Obviously, the difference between the classical and the quantum case is that, when we have an entanglement, we can have a perfect correlation between the apparatus and the object system, which is difficult to obtain in classical situations.

We would like now to show that the counterpart of the conditional probability $\wp(j|k)$ in quantum mechanics is indeed the probability of a final detection event, which, given a certain experimental context (a premeasurement), allows us to finally ascribe a property to the object system (the system that has been measured). Suppose that the initial state of the apparatus is some ready state $|A_0\rangle$ while the state of the object system is some superposition state $|\psi_S\rangle$. Then, the premeasurement step takes the form

$$|\psi_S\rangle |A_0\rangle \longrightarrow \sum_j c_j |j\rangle |a_j\rangle, \qquad (4.6)$$

where the $|j\rangle$s are the system states representing the eigenstates of some system observable to be measured and the $|a_j\rangle$s are respectively the apparatus states. In the density matrix formalism the initial state of the subject system and the apparatus appearing on the left-hand side of the transformation (4.6) may be described by the (factorized) density matrix $\hat{\rho}_S \hat{\rho}_A$, where

$$\hat{\rho}_S = |\psi_S\rangle\langle\psi_S| \quad \text{and} \quad \hat{\rho}_A = |A_0\rangle\langle A_0|. \qquad (4.7)$$

As usual we assume that the entanglement between the object system and the apparatus created during the premeasurement step is the result of a unitary transformation. Indeed, I remind the reader that only the final step of selection or detection is not unitary. Therefore, we have the following unitary transformation:

$$\hat{\rho}_S \hat{\rho}_A \mapsto \hat{U}_t \hat{\rho}_S \hat{\rho}_A \hat{U}_t^\dagger, \qquad (4.8)$$

where $\hat{U}_t$ is the time evolution operator whose form depends on the coupling of the system and the apparatus. It is easy to find that in the case under consideration we have

$$\hat{U}_t \hat{\rho}_S \hat{\rho}_A \hat{U}_t^\dagger = \sum_j |c_j|^2 |j\rangle\langle j| \otimes |a_j\rangle\langle a_j| + \sum_{j \neq k} c_j c_k^* |j\rangle\langle k| \otimes |a_j\rangle\langle a_k|. \qquad (4.9)$$

Just before the detection, the probability that the apparatus will read the value a_m is given by

$$\wp(a_m) = \text{Tr}_{\mathcal{A}}[\hat{P}_{a_m} \text{Tr}_S(\hat{U}_t \hat{\rho}_S \hat{\rho}_A \hat{U}_t^\dagger)], \qquad (4.10)$$

where $\hat{P}_{a_m} = |a_m\rangle\langle a_m|$ is the projector to the apparatus state $|a_m\rangle$. Note that the previous probability only takes into account correlated terms due to the entanglement between object system and apparatus. The above result can be verified quite easily. Indeed, the partial trace over the system will sum only the diagonal terms in the system Hilbert space $\mathcal{H}_S$, while the subsequent application of the projector $\hat{P}_{a_m}$ projects out a single term in the apparatus Hilbert space $\mathcal{H}_A$, yielding

$$\hat{P}_{a_m} \mathrm{Tr}_S(\hat{U}_t\hat{\rho}_S\hat{\rho}_A\hat{U}_t^\dagger) = |c_m|^2 |a_m\rangle\langle a_m|. \tag{4.11}$$

Finally, tracing out the apparatus, we shall get the probability, which, in our case, is simply given by $|c_m|^2$. Using the cyclic property of the trace and the fact that Tr_S does not act on the apparatus, we may rewrite Eq. (4.10) as

$$\begin{aligned}
\wp(a_m) &= \mathrm{Tr}_{\mathcal{A}}[\hat{P}_{a_m} \mathrm{Tr}_S(\hat{U}_t\hat{\rho}_S\hat{\rho}_A\hat{U}_t^\dagger)] \\
&= \mathrm{Tr}_{\mathcal{A}}[\mathrm{Tr}_S(\hat{U}_t^\dagger\hat{P}_{a_m}\hat{U}_t\hat{\rho}_S\hat{\rho}_A)] \\
&= \mathrm{Tr}_{\mathcal{A}}[\mathrm{Tr}_S(\hat{U}_t^\dagger|a_m\rangle\langle a_m|\hat{U}_t\hat{\rho}_S)\hat{\rho}_A].
\end{aligned} \tag{4.12}$$

It is convenient to define the Hermitian operator $\hat{E}_m$ in the system–apparatus Hilbert space by

$$\hat{E}_m = \mathrm{Tr}_S(\hat{U}_t^\dagger\hat{P}_{a_m}\hat{U}_t\hat{\rho}_S), \tag{4.13}$$

in terms of which the above equation can be written as

$$\wp(a_m) = \mathrm{Tr}_{\mathcal{A}}(\hat{E}_m\hat{\rho}_A). \tag{4.14}$$

The projection-like operator $\hat{E}_m$ is called the *effect operator* (or effect for short), which plays an important role in the theory of generalized measurement [18, 8]. From the definition (4.13), it follows that

$$\sum_j \hat{E}_j = \hat{I}, \tag{4.15}$$

where $\hat{I}$ is the identity operator (in the apparatus Hilbert space $\mathcal{H}_A$) and use has been made of the completeness relation for the apparatus states:

$$\sum_j \hat{P}_{a_j} = \hat{I}. \tag{4.16}$$

It can be shown that $\hat{E}_j$ is positive semidefinite, i.e.,

$$\langle\phi_A|\hat{E}_j|\phi_A\rangle \geq 0 \quad \text{for all} \quad |\phi_A\rangle \in \mathcal{H}_A. \tag{4.17}$$

However, unlike the projectors, the effect operators in general do not satisfy the requirement of orthogonality, that is,

$$\hat{E}_j\hat{E}_k \neq \delta_{jk}\hat{E}_k. \tag{4.18}$$

Moreover, substituting $\hat{\rho}_S = |\psi_S\rangle\langle\psi_S|$ into Eq. (4.13), we can explicitly calculate the trace over the system to obtain

$$
\begin{aligned}
\hat{E}_m &= \mathrm{Tr}_S(\hat{U}_t^\dagger |a_m\rangle\langle a_m|\hat{U}_t|\psi_S\rangle\langle\psi_S|) \\
&= \sum_{k=0,1} \langle k|\hat{U}_t^\dagger |a_m\rangle\langle a_m|\hat{U}_t|\psi_S\rangle \langle\psi_S|k\rangle \\
&= \langle\psi_S|\hat{U}_t^\dagger |a_m\rangle\langle a_m|\hat{U}_t|\psi_S\rangle \\
&= \hat{\vartheta}_m^\dagger \hat{\vartheta}_m,
\end{aligned}
\tag{4.19}
$$

where

$$
\hat{\vartheta}_m = \langle a_m|\hat{U}_t|\psi_S\rangle, \quad \hat{\vartheta}_m^\dagger = \langle\psi_S|\hat{U}_t^\dagger|a_m\rangle.
\tag{4.20}
$$

These expressions do *not* represent probability amplitudes because the time evolution operator $\hat{U}_t$ describes the coupling of the apparatus and the system, whereas the kets $|a_m\rangle$ and $|\psi_S\rangle$ represent respectively only the apparatus state and the system state. As a result, $\hat{\vartheta}_m$ is called the *amplitude operator*. From its definition, it is clear that the amplitude operator $\hat{\vartheta}_m$ describes the three steps of the measurement of a given observable:

(i) preparation of the initial state of the system (i.e., the input $|\psi_S\rangle$);

(ii) unitary time evolution (i.e., coupling or premeasurement) that entangles the system with the apparatus and allows us to select an observable (i.e., the processing represented by $\hat{U}_t$);

(iii) detection by the apparatus (i.e., the output $|a_m\rangle$) that allows us to assign a property to the system.

We can also summarize what is said here by writing

$$
\hat{\rho}_A^{(f)} = \frac{1}{\wp(a_m)}\hat{\vartheta}_m\hat{\rho}_A\hat{\vartheta}_m^\dagger,
\tag{4.21}
$$

where $\hat{\rho}_A^{(f)}$ is the state of the apparatus after the detection corresponding to the value a_m. Let the parameter m be associated with one of the detection outcomes a_m that corresponds to the apparatus state $|a_m\rangle$ and the parameter k with one of the state vectors $|\psi_k\rangle$ in a given orthonormal basis for the object system. Then, following a similar analysis as above and using Eqs. (4.14) and (4.19), we have

$$
\wp(m|k) = \mathrm{Tr}_{\mathscr{A}}(\hat{\vartheta}_{mk}\hat{\rho}_A\hat{\vartheta}_{mk}^\dagger),
\tag{4.22}
$$

where

$$
\hat{\vartheta}_{mk} = \langle a_m|\hat{U}_t|\psi_k\rangle.
\tag{4.23}
$$

Therefore, we have shown that the amplitude operator $\hat{\vartheta}_{mk}$ is the quantum mechanical counterpart of the classical conditional probability $\wp(j|k)$ given by Eq. (4.2).

This formalism also allows a new understanding of quantum mechanical formalism that fits very well with our treatment in terms of information [14]. We may take any ket $|\psi\rangle$ as the input state and any bra $\langle\varphi|$ as the output state. In this way, any projector $|\psi\rangle\langle\psi|$ is a selection act while any scalar product $\langle\varphi|\psi\rangle$ means a possible transition (and therefore any coefficient $c_\varphi = \langle\varphi|\psi\rangle$ means a transition amplitude). In the case in which it is equal to zero, it mean a forbidden transition (as a matter of fact any measurement deals with the collapse into one of the components of an initial superposition). If whenever $\langle\varphi_1|\psi\rangle = 0$ and $\langle\varphi_2|\psi\rangle = 0$ we also have $\langle\varphi|\psi\rangle = 0$, this means that $|\varphi\rangle$ is a coherent superposition of $|\varphi_1\rangle$ and $|\varphi_2\rangle$. Any operator that is in between an input and an output (like the above amplitude operators or, more generally, $\langle\varphi|\hat{O}|\psi\rangle$) is the operation that bridges between input and output. When input and output coincide, i.e. $\langle\psi|\hat{O}|\psi\rangle$, we have the probability amplitude that the transformation $\hat{O}$ does not change the state $|\psi\rangle$, that corresponds to the mean value of $\hat{O}$ in that state.

4.3 Quantum non-demolition measurement

When we are now interested in a measurement of the state vector, we have two distinct problems: (i) how to avoid the disruptive back action on the state of the measured system during detection and (ii) how to extract the complete information from this state. Actually, the two problems are interconnected, since in order to extract the complete information we need several measurement steps and it is mandatory to avoid the disruption of the state in each of these measurement steps. Now the question is: if we avoid back action, can we guarantee extraction of the whole information? These two different issues have been dealt with in two different areas of research: quantum non-demolition measurement and protective measurement, respectively. So, let us start with the first approach. *A quantum non-demolition* (QND) *measurement* is a measurement in which an apparatus extracts information only on the observable to be measured and transfers the whole back action on the canonical conjugate observable. In other words, the observable to be measured remains unperturbed, while the canonically conjugate one is perturbed precisely to the minimal extent allowed by the uncertainty relations. In order to examine the properties of a QND measurement, we need to introduce first the concept of indirect measurement [4, section 9.11][7].

4.3.1 Indirect measurement

We can treat this aspect by making use of the previous formalism. An indirect measurement is characterized by two different steps: first, a system S interacts with another quantum system S_P, the quantum probe, whose initial state has been accurately prepared on purpose in advance. This is an intermediate system with which the object system S interacts, and from which the apparatus $\mathcal{A}$ extracts information about S. During the first step there is no reduction at all, and the evolution is completely unitary, resulting in a correlation between S and S_P. In other words, the system and the probe become entangled. The second step consists of a direct measurement of some chosen observable of S_P: the state of the probe (and therefore, due to the entanglement, also of the object system) is reduced and the information acquired.

Let us now introduce two conditions referring to the two steps defined above:

- the reduction, i.e. the second step of the measurement, should begin only when the unitary evolution, i.e. the first step, has already finished;
- the second step should not contribute significantly to the total error of the measurement.

If these conditions are satisfied, we can infer the magnitudes of the error in the measurement and therefore of the perturbation (back action) of S from an analysis of the first step only, i.e. of the unitary evolution, because the only source of error is due to the intrinsic uncertainties of the initial state of S_P.

We can describe the indirect measurement in a formal way as follows. Let the *first step* be represented in analogy with transformation (4.8) by the transformation

$$\hat{\rho}_S \hat{\rho}_P \mapsto \hat{U}_t \hat{\rho}_S \hat{\rho}_P \hat{U}_t^\dagger , \qquad (4.24)$$

where $\hat{\rho}_S \hat{\rho}_P$ is the total density matrix of the system $S + S_P$, and $\hat{U}_t$ is the coupling unitary–evolution operator. The corresponding state of S_P alone, after the interaction, is given by the reduced density matrix

$$\hat{\bar{\rho}}_P = \mathrm{Tr}_S \left(\hat{U}_t \hat{\rho}_S \hat{\rho}_P \hat{U}_t^\dagger \right) . \qquad (4.25)$$

Suppose that we want to measure the observable $\hat{x}$ on S. Thanks to the entanglement, it is possible to achieve a one-to-one correspondence between the observable $\hat{x}$ of S and a carefully chosen observable of S_P, say $\hat{p}_x$. We can then perform a "direct" measurement of $\hat{p}_x$ on S_P. Since this measurement contributes negligibly to the experiment's overall error, we can idealize it as arbitrarily accurate. Then, we can infer from the value of $\hat{p}_x$ on S_P the value x_m of the observable $\hat{x}$ on S. Because of the one-to-one correspondence we can use $\hat{x}$ as a substitute for $\hat{p}_x$ and hence use x_m not only as the inferred value of $\hat{x}$ but also as the result of a measurement on S_P *itself*, that is, the associated eigenstate of the probe can be denoted by $|x_m\rangle$. In other words, it is a kind of EPR-like procedure. Just before the second step of the measurement, the probability distribution of the measured value x_m in analogy with the probability (4.12) is given by

$$\wp(x_{\mathrm{m}}) = \mathrm{Tr}_{S_{\mathrm{P}}} \left[\hat{P}_{x_{\mathrm{m}}} \hat{\bar{\rho}}_{\mathrm{P}} \right] = \mathrm{Tr}_{S_{\mathrm{P}}} \left[|x_{\mathrm{m}}\rangle\langle x_{\mathrm{m}}| \, \mathrm{Tr}_{S}(\hat{U}_t \hat{\rho}_{\mathrm{S}} \hat{\rho}_{\mathrm{P}} \hat{U}_t^{\dagger}) \right] , \qquad (4.26)$$

which, using the linearity and the cyclic property of the trace, can be rewritten as

$$\wp(x_{\mathrm{m}}) = \mathrm{Tr}_{S} \left[\hat{E}(x_{\mathrm{m}}) \hat{\rho}_{\mathrm{S}} \right] , \qquad (4.27)$$

where, in a analogy with Eq. (4.13), we have

$$\hat{E}(x_{\mathrm{m}}) = \mathrm{Tr}_{S_{\mathrm{P}}} \left[\hat{U}_t^{\dagger} |x_{\mathrm{m}}\rangle\langle x_{\mathrm{m}}| \hat{U}_t \hat{\rho}_{\mathrm{P}} \right] . \qquad (4.28)$$

The back action of the entire two-step measurement on S is embodied in the final state of the object system. In fact, the above considerations imply that such a normalized final state is

$$\hat{\rho}_{\mathrm{S}}^{(\mathrm{f})}(x_{\mathrm{m}}) = \frac{1}{\wp(x_{\mathrm{m}})} \langle x_{\mathrm{m}}| \hat{U}_t \hat{\rho}_{\mathrm{S}} \hat{\rho}_{\mathrm{P}} \hat{U}_t^{\dagger} |x_{\mathrm{m}}\rangle . \qquad (4.29)$$

If we expand the initial state of S_{P} as

$$\hat{\rho}_{\mathrm{P}} = \sum_k w_k |k\rangle\langle k| , \qquad (4.30)$$

where $|k\rangle\langle k|$ are some projectors on the probe's Hilbert space, then, by substituting this expression into Eq. (4.29), we obtain

$$\hat{\rho}_{\mathrm{S}}^{(\mathrm{f})}(x_{\mathrm{m}}) = \frac{1}{\wp(x_{\mathrm{m}})} \sum_k w_k \hat{\vartheta}_k(x_{\mathrm{m}}) \hat{\rho}_{\mathrm{S}} \hat{\vartheta}_k^{\dagger}(x_{\mathrm{m}}) , \qquad (4.31)$$

where, in analogy with Eqs. (4.20), we have

$$\hat{\vartheta}_k(x_{\mathrm{m}}) = \langle x_{\mathrm{m}}| \hat{U}_t |k\rangle . \qquad (4.32)$$

4.3.2 QND measurement

Now we can discuss the QND measurement. The central ingredient that makes the QND procedure realizable is just the two-step measurement process described previously [7]. It is then clear that some features of the indirect measurement also characterize the QND measurement. In general, we may say that, in a QND measurement, the system S interacts only with a probe S_{P}, and the interaction between S and S_{P} is such that S_{P} is influenced only by one observable, or a set of observables, that are not affected by the back action of S_{P} on S. More precisely, the system's observables which influence the probe must all commute with each other – i.e. they should belong to the same complete set of observables.

Moreover, a QND measurement can be performed only on observables that are conserved during the object's free evolution, i.e. on constants of motion. In the absence of external forces, the observable is conserved both during the measurement (because the back action is transferred to the canonically conjugate observable only) and during the unitary evolution between consecutive measurements (because it is an integral of motion). The above considerations imply that a QND measurement does not add any perturbation to the observable to be measured, so that the uncertainty of the measured observable after the measurement is only a consequence of the *a priori* uncertainty of its value.

Then, the observable $\hat{O}_{\text{ND}}$ associated with a QND measurement must satisfy two requirements:

- at any time it must commute with itself at a different time,

$$\left[\hat{O}_{\text{ND}}(t), \hat{O}_{\text{ND}}(t')\right] = 0, \quad t' \neq t, \tag{4.33}$$

- it must commute with the time-displacement unitary operator $\hat{U}_t$:

$$\left(\hat{U}_t^\dagger \hat{O}_{\text{ND}} \hat{U}_t - \hat{O}_{\text{ND}}\right)|\psi\rangle = 0, \tag{4.34}$$

where $|\psi\rangle$ is the probe's initial state and the expression between brackets is the Heisenberg-picture change in $\hat{O}_{\text{ND}}$ produced by the interaction between S and S_{P}. Equation (4.34) represents a necessary and sufficient condition of a QND observable.

We can therefore say that a QND measurement is characterized by the *repeatability*, so that the first measurement – which determines the values for all subsequent QND ones – is a preparation of S in the desired state, and the others are the determination of the value. As a consequence, we have

$$\hat{O}_{\text{ND}}(t_k) = f_k\left[\hat{O}_{\text{ND}}(t_0)\right], \tag{4.35}$$

where t_k is some arbitrary time after the initial t_0 (the time of the first measurement), and f_k is some real-valued function. Note that condition (4.35) implies condition (4.33).

4.3.3 No measurement without a measurement

We have seen that a QND measurement does not add any perturbation to the measured observable. This means that the standard deviation of the probability distribution of the measured observable is not altered by a QND measurement. One might think that the entire probability distribution of the measured observable is not altered by a QND measurement. If this were the case, then repeated QND measurements of the same observable could increase the amount of information which we can extract from the system. By repeating this procedure with different observables, we could extract complete information from the state vector of the target system. Here, I show that if a QND measurement performed on a system S does not alter the probability density of the measured observable, then the measurement process does not provide *any* information about the measured observable itself.

Let us start from the definition of the amplitude operator given by Eq. (4.32):

$$\hat{\vartheta}(x_m, \hat{x}_S) = \langle x_m | \hat{U}_t(\hat{x}_m, \hat{x}_S) | \psi_m \rangle \,, \tag{4.36}$$

where $\hat{x}_m$ is the measured observable of the probe (meter), x_m is its measured value, $\hat{x}_S$ is the (QND) observable of the system S, and $|\psi_m\rangle$ is the initial state of the apparatus. In Eq. (4.36) we have explicitly introduced the dependence of the amplitude operator, which completely describes the measurement, on $\hat{x}_S$ through the unitary operator $\hat{U}_t$.

The QND condition for a back-action-evading measurement then means that $\hat{x}_S$ and $\hat{\vartheta}$ share the same eigenstates:

$$\hat{\vartheta}(\hat{x}_S, x_m) | x_S \rangle = \vartheta(x_m, x_S) | x_S \rangle \,, \tag{4.37a}$$

$$\hat{\vartheta}^\dagger(\hat{x}_S, x_m) | x_S \rangle = \vartheta^*(x_S, x_m) | x_S \rangle \,. \tag{4.37b}$$

After a measurement on the meter which gives the result x_m, in analogy with Eq. (4.21) the system is described by the density matrix

$$\hat{\rho}_S^{(f)}(x_m) = \frac{1}{\wp(x_m)} \hat{\vartheta}(x_S, x_m) \hat{\rho}_S \hat{\vartheta}^\dagger(x_S, x_m) \,, \tag{4.38}$$

where $\wp(x_m)$ may be written as

$$\wp(x_m) = \text{Tr}_S \left[\hat{\vartheta}(\hat{x}_S, x_m) \hat{\rho}_S \hat{\vartheta}^\dagger(\hat{x}_S, x_m) \right]$$
$$= \int dx_S \, \langle x_S | \hat{\vartheta}(\hat{x}_S, x_m) \hat{\rho}_S \hat{\vartheta}^\dagger(\hat{x}_S, x_m) | x_S \rangle \,. \tag{4.39}$$

Now, the probability density of the measured observable after the measurement is given by

$$\wp_f(x_S) = \langle x_S | \hat{\rho}_S^{(f)}(x_m) | x_S \rangle$$
$$= \frac{1}{\wp(x_m)} \langle x_S | \hat{\vartheta}(\hat{x}_S, x_m) \hat{\rho}_S \hat{\vartheta}^\dagger(\hat{x}_S, x_m) | x_S \rangle \,. \tag{4.40}$$

Applying the QND conditions (4.37a) we obtain

$$\wp_f(x_S) = \frac{1}{\wp(x_m)} |\vartheta(x_S, x_m)|^2 \, \wp(x_S) \,, \tag{4.41}$$

where $\wp(x_S)$ is the initial state's *a priori* probability distribution of $\hat{x}_S$, given by

$$\wp(x_S) = \text{Tr} \left[\hat{E}(x_m) \hat{\rho}_S \right]. \tag{4.42}$$

If we require that the probability density (4.41) does not change due to the measurement process, i.e. $\wp_f(x_S) = \wp(x_S)$, then

$$|\vartheta(x_S, x_m)|^2 = \wp(x_m) \tag{4.43}$$

must hold. However, $\wp(x_m)$ is not a function of x_S (the eigenvalues of the measured observable) and therefore also the eigenvalues $\vartheta(x_S, x_m)$ of $\hat{\vartheta}(\hat{x}_S, x_m)$ are independent of x_S. Since the operator $\hat{\vartheta}$ must completely describe the measurement process, if its eigenvalues are independent of the eigenvalues of $\hat{x}_S$, the measurement obviously gives no information about $\hat{x}_S$, unless the initial state is already an eigenstate of the measured observable. In conclusion, evading the back-action does not guarantee the necessary accumulation of information that would allow us to speak of measurement of the state vector. Are there other ways to address this problem?

4.4 Protective measurement of the state

It is Aharonov and co-workers [1, 2] (see also [4, section 15.1]) who have tried this path. The main idea is that of *protective measurement*, i.e. a measurement of the wave function during which it is prevented from changing noticeably by means of another interaction which it undergoes at the same time. The reason for this approach is the following: since the wave function is an elusive entity and, as we have seen, any attempt at measuring a quantum system will in general alter the initial state, a protective measurement could ensure us the preservation of the state allowing us to simultaneously extract some information about it.

Let us assume that we wish to measure an observable $\hat{O}$ on a system in the state $|\varsigma\rangle = \sum c_j |o_j\rangle$, where the states $|o_j\rangle$ are eigenkets of $\hat{O}$, so that

$$\hat{O}|o_j\rangle = o_j|o_j\rangle , \tag{4.44}$$

and that the interaction between the apparatus $\mathcal{A}$ and the system $\mathcal{S}$ is a part of the total Hamiltonian by the Hamiltonian

$$\hat{H}_{\mathcal{A}+\mathcal{S}} = \hat{H}_0 + \hat{H}_{\mathcal{A}\mathcal{S}} + \hat{H}_{\mathcal{A}} , \tag{4.45}$$

where

$$\hat{H}_{\mathcal{A}\mathcal{S}} = \hat{H} = \varepsilon(t)\hat{x}_{\mathcal{A}}\hat{O} \tag{4.46}$$

is the interaction Hamiltonian, $\hat{H}_0$ is the free Hamiltonian of the system, $\hat{H}_{\mathcal{A}}$ is the Hamiltonian of the apparatus, $\hat{x}_{\mathcal{A}}$ is the one-dimensional pointer observable, and ε represents the coupling function, i.e. $\varepsilon(t)$ is non-zero only in the interval $[0, \tau]$ (duration of the interaction). In general, such an interaction leads to an entangled state, which may be written as

$$|\Psi(\tau)\rangle = \sum_j c'_j|o_j\rangle|a_j\rangle , \tag{4.47}$$

where

$$|a_j\rangle = e^{-(i/\hbar)\varepsilon o_j \hat{x}_{\mathcal{A}}}|A_0\rangle \tag{4.48}$$

are states of the apparatus $\mathcal{A}$ which, for sufficiently large ε, are orthogonal for distinct eigenvalues a_j of the pointer observable. The apparatus $\mathcal{A}$ is in the initial state $|A_0\rangle$.

Let us now consider the case in which no entanglement takes place. Therefore, in the place of Eq. (4.47), we write the factorized state

$$|\varsigma(0)\rangle|A(0)\rangle \mapsto |\varsigma(t)\rangle|A(t)\rangle, \ t > 0 . \tag{4.49}$$

In this case, there is no reduction of the wave function. Instead, we would have the equation of motion

$$\frac{d}{dt}\langle\varsigma(t)|\langle A(t)|\hat{p}_x^{\mathcal{A}}|\varsigma(t)\rangle|A(t)\rangle = -\varepsilon(t)\langle\varsigma(t)|\hat{O}|\varsigma(t)\rangle, \tag{4.50}$$

where $\hat{p}_x^{\mathcal{A}}$ is the observable canonically conjugate to $\hat{x}_{\mathcal{A}}$ and, in the Heisenberg picture,

$$\frac{d}{dt}\hat{p}_x^{\mathcal{A}} = \frac{i}{\hbar}\left[\hat{H}, \hat{p}_x^{\mathcal{A}}\right] = -\varepsilon(t)\hat{O} , \tag{4.51}$$

where the commutation relations position–momentum and Eq. (4.46) have been used. Equation (4.51) shows that $\hat{p}_x^{\mathcal{A}}$ changes by different amounts for distinct eigenvalues o_j, and by Eq. (4.50) we can determine $\langle\varsigma(t)|\hat{O}|\varsigma(t)\rangle$ by the change in the apparatus' momentum.

A protective measurement can be made in two different ways.

(i) If $|\varsigma(t)\rangle$ is a non-degenerate eigenstate of the Hamiltonian $\hat{H}$, then the interaction is assumed to be sufficiently weak and $\hat{H}$ changes slowly so that $|\varsigma(t)\rangle$ is nearly equal to $|\varsigma(0)\rangle$ up to a phase factor for $t \in [0,\tau]$. Then, following the adiabatic theorem, $|\varsigma(t)\rangle$ remains an eigenstate of the Hamiltonian and no entanglement takes place.

(ii) If we have an arbitrary evolution, so that $|\varsigma(t)\rangle$ is not necessarily an eigenstate of the Hamiltonian, we can operate in the following manner. If $|\varsigma_0(t)\rangle$ is the evolution of $|\varsigma\rangle$ determined by the unperturbed Hamiltonian $\hat{H}_0$ of the system $\mathcal{S}$, then one can measure an observable $\hat{O}'(t)$, for which $|\varsigma_0(t)\rangle$ is a non-degenerate eigenstate, a large number of times which are dense in the interval $[0,\tau]$ – say at times $t_n = (n/N)\tau, n = 1, 2, \ldots, N$, where N is an arbitrarily large number. Then, $|\varsigma(t)\rangle$ does not noticeably depart from $|\varsigma_0(t)\rangle$ – it is a sort of quantum Zeno effect [9, 19]. Now consider the branch of the combined system evolution in which each measurement of $\hat{O}'(t_n)$ results in the state $|\varsigma_0(t_n)\rangle$ of $\mathcal{S}$:

$$|\Psi(\tau)\rangle_0 = \prod_N \hat{P}_{0,N} \hat{U}_{t_N} |A(0)\rangle = \prod_N \hat{P}_{0,N} \hat{U}_{t_N,\varepsilon} |A_0(\tau)\rangle, \qquad (4.52)$$

where

$$\hat{P}_{0,N} = |\varsigma_0(t_N)\rangle\langle\varsigma_0(t_N)|, \qquad (4.53)$$

$$\hat{U}_{t_N} = e^{-\frac{i}{\hbar}\frac{\tau}{N}\hat{H}(t_N)}, \qquad (4.54)$$

$$\hat{U}_{t_N,A} = e^{-\frac{i}{\hbar}\frac{\tau}{N}\varepsilon(t_N)\hat{x}_A\hat{O}}, \qquad (4.55)$$

and $|A_0(\tau)\rangle$ is the state of $\mathcal{A}$ when it evolves under the Hamiltonian $\hat{H}_{\mathcal{A}}$. We now calculate explicitly the last expectation value in Eq. (4.52) up to second order in $1/N$ and find

$$
\begin{aligned}
\langle\varsigma_0(t_1)|e^{-\frac{i}{\hbar}\frac{\tau}{N}\varepsilon(t_1)\hat{x}_A\hat{O}}|\varsigma_0(t_1)\rangle &= 1 - \frac{i}{\hbar}\frac{\tau}{N}\varepsilon(t_1)\hat{x}_A\langle\hat{O}\rangle - \frac{1}{2\hbar^2}\frac{\tau^2}{N^2}\varepsilon(t_1)^2\hat{x}_A^2\langle\hat{O}^2\rangle \\
&= 1 - \frac{i}{\hbar}\frac{\tau}{N}\varepsilon(t_1)\hat{x}_A\langle\hat{O}\rangle \\
&\quad - \frac{1}{2\hbar^2}\frac{\tau^2}{N^2}\varepsilon(t_1)^2\hat{x}_A^2\langle\hat{O}\rangle^2 - \frac{1}{2\hbar^2}\frac{\tau^2}{N^2}\varepsilon(t_1)^2\hat{x}_A^2\Delta\hat{O}^2 \\
&= e^{-\frac{i}{\hbar}\frac{\tau}{N}\varepsilon(t_1)\hat{x}_A\langle\hat{O}\rangle}\left[1 - \frac{1}{2\hbar^2}\frac{\tau^2}{N^2}\varepsilon(t_1)^2\hat{x}_A^2\Delta\hat{O}^2\right],
\end{aligned}
$$

$$(4.56)$$

where we have made use of the fact that

$$\Delta\hat{O}^2 = \langle\hat{O}^2\rangle - \langle\hat{O}\rangle^2. \qquad (4.57)$$

In the limit $N \rightarrow \infty$, where the product of the factors in the term containing $\Delta\hat{O}^2$ approaches 1, Eq. (4.52) reads

$$|\Psi(\tau)\rangle_0 = |\varsigma_0(\tau)\rangle \exp\left(-\frac{i}{\hbar}\int_0^\tau dt\varepsilon(t)\hat{x}_A\langle\hat{O}\rangle\right)|A_0(\tau)\rangle. \qquad (4.58)$$

In this limit, the considered branch undergoes a unitary evolution and therefore the contribution from other branches – giving rise to states different from $|\varsigma_0(t)\rangle$ – vanishes. From the exponential operator in Eq. (4.58), the momentum of the apparatus is shifted by an amount (see also Eq. (4.51))

$$\Delta\hat{p}_x^{\mathcal{A}} = -\int_0^\tau dt\langle\hat{O}\rangle\varepsilon(t). \qquad (4.59)$$

Therefore, by measuring $\hat{p}_x^{\mathcal{A}}$, $\langle \hat{O} \rangle$ can be determined. Then, according to the present approach, by repeating this experiment with different observables, the wave function of a single system may be determined up to an overall phase factor. Moreover, since a protective measurement as proposed by Aharonov and coworkers should not give rise to entanglement between the system and the apparatus nor lead to a collapse, it could allow us in principle to distinguish between two non-orthogonal states, provided that both are protected.

The question naturally arises whether this proposal aimed at measuring the state of a quantum system implies that we should treat it as an observable. In other words, does this approach imply that we should consider the quantum state in classical terms? Let us suppose that this is the case. We know that the density matrix describing a pure state is a projector like $\hat{P}_{0,N}$, i.e. it is an observable. So, why one can measure a projector but should not obtain information about the state? The question is: what are the possible values that we would obtain by measuring a projector? Obviously, 0 or 1. If we obtain 0, we know that the system has not passed a certain test (say a vertical polarization filter), whereas if we obtain 1, we know that it has passed it. However, if the system before the test was in a superposition state of, say, vertical and horizontal polarization, we have a non-zero probability that it passes and a non-zero probability that it does not pass the test. Therefore, if we obtain a 0, we are not able to distinguish whether the system before the measurement was in a horizontal polarization state or in a superposition of vertical and horizontal polarization, and, similarly, if we obtain 1, we cannot distinguish between a previous vertical or superposed polarization state. In conclusion, the measurement of a projector (which, of course, is always possible) is not able to discriminate between non-orthogonal states. In other words, given an unknown state, we cannot decide *which* projector, if measured, would allow us to determine it. So, it does not seem that we can discriminate among non-orthogonal states by measuring projectors.

4.5 Measurement and reversibility

If through a protective measurement we can avoid a collapse and keep the dynamics of the system as ruled by unitary operators (or reversible operations), the natural question arises whether we can extract any information from a system when performing a reversible measurement. Nielsen and Caves [20] have shown that this is not possible. We have a unitarily reversible measurement – on a subspace $\mathcal{H}_0$ of the state space $\mathcal{H}$ of the original problem – if there exists a unitary operator $\hat{U}$ acting on $\mathcal{H}_0$ such that

$$\hat{\rho}_S^{(f)} = \hat{U} \frac{\vartheta \hat{\rho}_S \vartheta^\dagger}{\mathrm{Tr}[\vartheta \hat{\rho}_S \vartheta^\dagger]} \hat{U}^\dagger \tag{4.60}$$

for all $\hat{\rho}_S$ whose support lies in $\mathcal{H}_0$. This means that the POVM $\hat{E} = \hat{\vartheta}^\dagger \hat{\vartheta}$, when restricted to $\mathcal{H}_0$, is a positive multiple of the identity operator on $\mathcal{H}_0$, i.e.

$$\hat{P}_{\mathcal{H}_0} \hat{E} \hat{P}_{\mathcal{H}_0} = \eta^2 \hat{P}_{\mathcal{H}_0} , \qquad (4.61)$$

where η is a real constant satisfying $0 < \eta \le 1$ and $\hat{P}_{\mathcal{H}_0}$ is the projector onto $\mathcal{H}_0$. In other words, we have

$$\hat{P}_{\mathcal{H}_0} + \hat{P}_{\mathcal{H}_0^\perp} = \hat{I} , \qquad (4.62)$$

where $\hat{P}_{\mathcal{H}_0^\perp}$ projects onto the subspace that is complementary to $\mathcal{H}_0$. Then, the operator $\hat{\vartheta}$ can be written as

$$\hat{\vartheta} = \eta \hat{U} \hat{P}_{\mathcal{H}_0} + \hat{\vartheta} \hat{P}_{\mathcal{H}_0^\perp} , \qquad (4.63)$$

where $\hat{U}$ is some unitary operator acting on the whole $\mathcal{H}$. The conclusion, however, is that we have

$$\text{Tr}\left[\hat{\rho}_S \hat{\vartheta}^\dagger \hat{\vartheta} \right] = \text{Tr}\left[\hat{\rho}_S \hat{E} \right] = \eta^2 \qquad (4.64)$$

for all density operators whose support lies in $\mathcal{H}_0$, and η^2 takes the meaning of the probability of occurrence of the result represented by $\hat{\vartheta}$. In other words, the probability of occurrence of any measurement result represented by ϑ is the same for all states $|\psi\rangle$ (such that $\hat{\rho} = |\psi\rangle\langle\psi|$) that are normalized in that subspace.

Summarizing, an ideal measurement is reversible if and only if no new information about the prior state is obtained from the measurement. Given the result as stated by Eq. (4.64), each state is equally likely. As a consequence, a reversible ideal measurement cannot be considered as a true measurement.

4.6 Quantum state reconstruction

The conclusion of the previous section seems very unsatisfactory for protective measurement. Nevertheless, the idea that we can reconstruct the state of a quantum system was promising. However, we need also to renounce considering a quantum state in classical terms, as I show now. In this section, I discuss a few methods which allow the reconstruction of the quantum state on a large set of identical systems. Royer [22, 23] analyzed the problem in general terms. The problem can be cast as follows. Given a well-defined preparation procedure and a certain number of identical systems, is it possible to determine experimentally (to measure) the state which such a procedure forces the systems to be in? Due to the one-to-one correspondence between the Wigner function (W-function) and the density matrix of a system, this is possible if one is able to determine the W-function. Indeed, there are circumstances where a direct measurement of the W-function is simply more convenient. In other words, the measurement of the Wigner function or any other of the methods discussed in the present section are only possible if one performs a large number of measurements, each one on a single element of a set of identically prepared systems.

Let us first make use of the following formalism. Generally speaking, any operator $\hat{O}$ may be represented in the following form:

$$\hat{O} = \sum_{j,k} |j\rangle\langle j|\hat{O}|k\rangle\langle k| \,, \qquad (4.65)$$

where $\{|n\rangle\}$ is an arbitrary basis on the underlying Hilbert space. In this way, operators may be considered as – not necessarily normalized – 'vectors' in a super Hilbert space, which is the direct product of the original Hilbert space $\mathcal{H}$ and its dual $\mathcal{H}^*$. In fact, Eq. (4.65) may be rewritten as

$$\hat{O} = \sum_{j,k} O_{jk}|j\rangle\langle k| \,, \qquad (4.66)$$

where

$$O_{jk} = \langle j|\hat{O}|k\rangle \,. \qquad (4.67)$$

Therefore, we may associate with any operator $\hat{O}$ an S-ket $|\hat{O}\}$ and an S-bra $\{\hat{O}|$, defined by

$$|\hat{O}\} = \sum_{j,k} O_{jk}|j,k\} \,, \qquad (4.68a)$$

$$\{\hat{O}| = \sum_{j,k} O_{jk}^*\{j,k| \,, \qquad (4.68b)$$

where

$$|j,k\} = \||j\rangle\langle k|\} \,, \qquad (4.69a)$$

$$\{j,k| = \{|j\rangle\langle k|\| \qquad (4.69b)$$

represent the basis in which the S-ket (and the S-bra) is expanded. Their scalar product may be represented as

$$\{l,m|j,k\} = \langle l|j\rangle\langle k|m\rangle = \delta_{l,j}\delta_{k,m} \,, \qquad (4.70)$$

from which the generalized scalar product follows:

$$\{\hat{O}|\hat{O}'\} = \mathrm{Tr}(\hat{O}^\dagger \hat{O}') \,. \qquad (4.71)$$

From Eq. (4.71) it follows that we may reformulate Eq. (4.67) as

$$\langle j|\hat{O}|k\rangle = \{j,k|\hat{O}\} \,. \qquad (4.72)$$

As a consequence of the introduction of S-kets and S-bras, the operators acting on the S-kets and S-bras are called *superoperators*.

Let us limit ourselves to a one-dimensional system. whose phase state is represented by position $\hat{x}$ and momentum $\hat{p}_x$ [4, section 15.4]. Let us define the parity operator about the origin:

$$\hat{\Pi} = \int_{-\infty}^{+\infty} dx| - x\rangle\langle x| = \int_{-\infty}^{+\infty} dp_x| - p_x\rangle\langle p_x| ,\qquad (4.73)$$

where $-\infty < x < +\infty, -\infty < p_x < +\infty$. We may build the operator $\hat{\Pi}_{xp}$ about the phase-space point (x, p_x) by making use of the displacement operator

$$\hat{D}_{xp} = e^{\frac{i}{\hbar}(p_x\hat{x}-x\hat{p}_x)} ,\qquad (4.74)$$

which helps us to write

$$\hat{\Pi}_{xp} = \hat{D}_{xp}\hat{\Pi}\hat{D}_{xp}^{-1} ,\qquad (4.75)$$

where

$$\begin{aligned}
\hat{\Pi}_{xp} &= \frac{\hbar}{2} \int_{-\infty}^{+\infty} dx' e^{ix'p_x} \left|x + \frac{1}{2}\hbar x'\right\rangle \left\langle x - \frac{1}{2}\hbar x'\right| \\
&= \frac{\hbar}{2} \int_{-\infty}^{+\infty} dp'_x e^{ixp'_x} \left|p_x + \frac{1}{2}\hbar p'_x\right\rangle \left\langle p_x - \frac{1}{2}\hbar p'_x\right| \\
&= \frac{\hbar}{4\pi} \int_{-\infty}^{+\infty} dp'_x \int_{-\infty}^{+\infty} dx' e^{ip'_x(\hat{x}-x)-ix'(\hat{p}_x-p_x)} ,
\end{aligned}\qquad (4.76)$$

$\left|x + \frac{1}{2}\hbar x'\right\rangle$ and $\left|p_x + \frac{1}{2}\hbar p'_x\right\rangle$ being eigenkets of position and momentum, respectively. It follows indeed that

$$\hat{\Pi}_{xp}(\hat{x} - x)\hat{\Pi}_{xp} = -(\hat{x} - x), \hat{\Pi}_{xp}(\hat{p}_x - p_x)\hat{\Pi}_{xp} = -(\hat{p}_x - p_x) ,\qquad (4.77)$$

that is, $\hat{\Pi}_{xp}$ is the parity operator about the phase-space point (x, p_x). This allows me to introduce the S-vectors

$$|\hat{x}\hat{p}_x\} = \sqrt{\frac{2}{\pi\hbar}}|\hat{\Pi}_{xp}\} .\qquad (4.78)$$

The key point of the following discussion is that the Wigner function

$$W(x, p_x) = \frac{1}{\pi h} \int_{\mathfrak{R}} dx' \langle x + x'|\hat{\rho}|x - x'\rangle e^{2i\frac{p_x x'}{\hbar}} ,\qquad (4.79)$$

where $\hat{\rho}$ describes the state of the system under observation, is the expectation value of the parity operator $\hat{\Pi}_{xp}$ (see Eq. (4.71)):

$$W_{\hat{\rho}}(x, p_x, t) = \left\{\hat{\Pi}_{xp}|\hat{\rho}(t)\right\} = \frac{1}{\pi\hbar}\left\langle\hat{\Pi}_{xp}\right\rangle_{\hat{\rho}(t)}, \qquad (4.80)$$

where $\hat{\Pi}_{xp}$ is Hermitian. Since obviously $\hat{\Pi}_{xp}^2 = \hat{I}$, $\hat{\Pi}_{xp}$ is an observable whose eigenvalues are ± 1. A complete set of eigenstates $|\psi_{xp}^n\rangle$, $n = 1, 2, \ldots$, satisfying

$$\hat{\Pi}_{xp}|\psi_{xp}^n\rangle = (-1)^n|\psi_{xp}^n\rangle, \qquad (4.81)$$

may be obtained by displacing in phase space any complete orthogonal set of kets $|\psi^n\rangle$ of definite parity about the origin:

$$|\psi_{xp}^n\rangle = \hat{D}_{xp}|\psi^n\rangle, \qquad (4.82a)$$

$$\psi^n(-x) = (-1)^n\psi^n(x). \qquad (4.82b)$$

An immediate consequence of Eq. (4.81) is

$$\hat{\Pi}_{xp} = \sum_n (-1)^n|\psi_{xp}^n\rangle\langle\psi_{xp}^n|. \qquad (4.83)$$

This formalism allows us to rewrite Eq. (4.80) as

$$W_{\hat{\rho}}(x, p_x, t) = \frac{1}{\pi\hbar}\sum_n (-1)^n\langle\psi_{xp}^n|\hat{\rho}(t)|\psi_{xp}^n\rangle. \qquad (4.84)$$

We try now to measure $W_{\hat{\rho}}(x, p_x, t)$ at some definite time (e.g. $t = 0$). This can be done by measuring each transition probability $\langle\psi_{xp}^n|\hat{\rho}(0)|\psi_{xp}^n\rangle$ following the method introduced by Lamb. A simple approach is possible if we choose the $|\psi^n\rangle$s to be eigenstates of the Hamiltonian

$$\hat{H} = \frac{\hat{p}_x^2}{2m} + V(\hat{x}), \qquad (4.85)$$

where $V(-x) = V(x)$ is a symmetric potential. Then, the $|\psi_{xp}^n\rangle$s are eigenstates of the displaced Hamiltonian

$$\hat{H}_{xp} = \hat{D}_{xp}\hat{H}\hat{D}_{xp}^{-1} = \frac{(\hat{p}_x - p_x)^2}{2m} + V(\hat{x} - x), \qquad (4.86)$$

so that measuring the set $\langle\psi_{xp}^n|\hat{\rho}(0)|\psi_{xp}^n\rangle$ (or $\hat{\Pi}_{xp}$) becomes equivalent to measuring the Hamiltonian $\hat{H}_{xp}$. A suitable method to measure $\hat{H}_{xp}$ almost in the strict sense is as follows. First, we place ourselves in a reference frame moving with uniform speed $v = p_x^f/m$ relative to the preparation apparatus $\mathcal{A}$. By virtue of the Galilei transformations, the observed density operator (for $t \leq 0$) is

$$\hat{\rho}^f(t) = \hat{D}_{vt,p_x^f}^{-1}\hat{\rho}(t)\hat{D}_{vt,p_x^f}. \qquad (4.87)$$

At time $t = 0$ we turn on the potential $V(x - x^f)$ in the moving frame. The eigenstates of

$$\hat{H}_{x^f,0} = \frac{\left(\hat{p}_x^f\right)^2}{2m} + V(\hat{x}^f - x^f) \tag{4.88}$$

are

$$\hat{D}_{x^f,0}|\psi^n\rangle = |\psi^n_{x^f,0}\rangle, \tag{4.89}$$

with corresponding energies E_n. Then, at times $t \geq 0$ we obtain

$$\begin{aligned}
\hat{\rho}^f(t) &= e^{-\frac{i}{\hbar}t\hat{H}_{x^f,0}}\hat{\rho}^f(0)e^{\frac{i}{\hbar}t\hat{H}_{x^f,0}} \\
&= \sum_{m,n} e^{-\frac{i}{\hbar}(E_n-E_m)t}|\psi^n_{x^f,0}\rangle\langle\psi^m_{x^f,0}|\langle\psi^n_{x^f,0}|\hat{\rho}^f(0)|\psi^m_{x^f,0}\rangle \\
&= \sum_{m,n} e^{-\frac{i}{\hbar}(E_n-E_m)t}|\psi^n_{x^f,0}\rangle\langle\psi^m_{x^f,0}|\langle\psi^n_{x^f,p_x^f}|\hat{\rho}(0)|\psi^m_{x^f,p_x^f}\rangle, \tag{4.90}
\end{aligned}$$

where we have made use of the transformation (4.87) for $\hat{\rho}(0)$. Now, the transition probabilities

$$\langle\psi^n_{x^f,0}|\hat{\rho}^f(0)|\psi^n_{x^f,0}\rangle = \langle\psi^n_{x^f p_x^f}|\hat{\rho}(0)|\psi^n_{x^f p_x^f}\rangle \tag{4.91}$$

are time independent, so that we have a long time available to perform a measurement of $\hat{H}_{x^f,0}$ referring to the set $\{|\psi^n_{x^f,0}\rangle\}$ and "find" the particle in one of the states pertaining to this set. Repeating the measurement many times will allow us to build the distribution (4.91), from which $W_{\hat{\rho}}(x^f, p_x^f, 0)$ can be deduced by means of Eq. (4.84). What has been done is a measurement of $W_{\hat{\rho}}(x^f, p_x^f, 0)$ by measuring

$$W_{\hat{\rho}^f}(x^f, 0, t) = W_{\hat{\rho}}(x^f + vt, p_x^f, t) \tag{4.92}$$

at $t = 0$ in the moving frame. In conclusion, applying the same procedure over and over again with different values of x^f and p_x^f, it is in principle possible to reconstruct the Wigner function on any relevant region of the phase space.

What we have learnt is that any reconstruction of the state requires statistical methods applied to several systems prepared in the same state. In other words, we cannot extract this information from the measurement of a single system. Moreover, the information that we can extract by using statistical methods is never the whole information potentially contained in the initial state of the systems. This raises the issue whether and eventually in which conditions we can speak of informational completeness. Moreover, the Wigner function (and any other joint distribution of quantum conjugate observables) presents interference terms that affect the way in which we can consider the state. The latter two problems will be the object of the next section.

4.7 Unsharpness and negative quasi-probabilities

We shall consider now two further prices to pay for reconstructing the wave function of a quantum system, in terms of both:

- unsharpness of the observables and
- negative quasi-probabilities.

The problem of quantum state reconstruction raises the question of the extent to which an observable's measurement informs us about the state of a given system. Stated in other terms, we may ask ourselves whether the probability distributions of a certain set of observables are sufficient to determine the state of a quantum system, i.e. to discriminate between different states. Such a question leads naturally to the concept of *informational completeness* [15, 21]. A family of self-adjoint operators $\{\hat{O}_k\}$ is said to be informationally complete if

$$\text{Tr}\left[\hat{\rho}\hat{O}_k\right] = \text{Tr}\left[\hat{\rho}'\hat{O}_k\right] , \forall k, \tag{4.93}$$

implies that $\hat{\rho} = \hat{\rho}'$ on a Hilbert space $\mathcal{H}$. It is possible to show that sharp observables are not informationally complete.

We recall, instead, that unsharp observables are the result of a smearing operation on sharp ones [8, 10, 11, 12]. Now, this smearing operation on sharp observables, say $\hat{\mathbf{p}}$ and $\hat{\mathbf{r}}$, can be understood as a coarse-graining operation on a set of sharp observables. This operation can have an informationally complete refinement as a result.

About the issue of quasi-probabilities, we can see from Fig. 4.1 that the Wigner function can assume negative values, which is classically not allowed. In fact, this would imply a kind of negative probability [13, 24]. It is true that we can make use of these negative probabilities only for conditional probabilities and intermediate steps in order to ensure that the final probabilities of events are positive. Nevertheless, we can at most say that we deal with negative quasi-probabilities. Moreover, if we do and, at the same time, we want to assign an ontological status to the quantum state, the latter could not be of a classical kind. Resuming, in order to reconstruct the wave function of a quantum system we need to consider both conjugate observables like position and momentum. However, since these observables do not commute, which implies that relative to at least one of them the state is a superposition, there are interference terms, and those interference terms determine that the joint function can assume negative values. This leads us to admit negative quasi-probabilities, which implies that the quantum state cannot be considered in classical terms.

Both problems (the use of unsharp observables for dealing with informational completeness and the necessity to involve negative quasi-probabilities) have a common root that has been called *quantum features* [6], i.e. those non-local, specifically

Gennaro Auletta

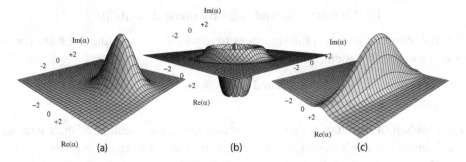

Figure 4.1 (a) Representation of the W-function of a coherent state $|\alpha\rangle = |2\rangle$. It is a bidimensional Gaussian centered at the point $\alpha_0 = (2,0)$ in the complex plane $(\Re(\alpha), \Im(\alpha))$. (b) Representation of the W-function of a number state $|n\rangle$ with $n = 4$, i.e. with the number of photons equal to the mean number of photons in the coherent state (a). Its annular shape shows the phase-invariance of the number state. Note that there are regions where the function becomes negative. (c) W-function representation of a squeezed state $|\alpha, \xi\rangle$ with $\alpha = 2$ and where $\xi = 0.8$ is the squeezing factor.

quantum interdependencies that are present in every quantum phenomenon and that in entanglement assume the form of quantum discord [27]. What is sad is that such features, although contributing to the probabilities of final detection events, can never be directly detected but can be experimentally ascertained by making use of the resources so far mentioned.

4.8 Conclusion

What we have learnt is that we cannot have direct evidence of, i.e. directly measure, a quantum state of a single system. Our experience is only connected with the experimental values of observables, and any time we measure an observable we can only make a partial experience of a system under a certain perspective but we can never have a complete experience that would be represented by an observation of the state vector. In other words, the quantum state is *not* an observable in the classical sense. However, since this feature of the quantum state is not due to subjective ignorance but rather to an intrinsic characteristic of the microscopic world, there are no definitive reasons to deny the reality of a quantum state. In fact, we need to admit that the quantum state is intrinsically affected by non-local features as expressed by the interference terms [6] and those terms forbid both a classical interpretation of the state and a classical measurement of the latter.

Given these provisos, it makes perfect sense to deal with the measurement of the quantum state, provided that we:

- make use of statistical methods;
- accept the partiality of our information acquiring (and therefore make use of generalized POVMs or unsharp observables);
- accept the consequence of quantum interference terms, i.e. negative quasi-probabilities.

I think that future developments of protective measurement satisfying these requirements could be very promising.

References

[1] Aharonov, Y. and Vaidman, L., Measurement of the Schrödinger wave of a single particle, *Physics Letters A* **178**: 38–42 (1993).

[2] Aharonov, Y., Anandan, J., and Vaidman, L., Meaning of the wave function, *Physical Review* **A47**: 4616–4626 (1993).

[3] Auletta, G., *Foundations and Interpretation of Quantum Mechanics; in the Light of a Critical-Historical Analysis of the Problems and of a Synthesis of the Results*, Singapore, World Scientific (2000; rev. edn. 2001).

[4] Auletta, G., Fortunato, M., and Parisi, G., *Quantum Mechanics*, Cambridge, Cambridge University Press (2009, 2014).

[5] Auletta, G., *Cognitive Biology: Dealing with Information from Bacteria to Minds*. Oxford, Oxford University Press (2011).

[6] Auletta, G. and Torcal, L., From wave-particle to features-event complementarity, *International Journal of Theoretical Physics* **50**: 3654–3668 (2011).

[7] Braginsky, V. B. and Khalili, F. Y., *Quantum Measurement*, Cambridge, Cambridge University Press (1992).

[8] Busch, P., Grabowski, M., and Lahti, P. J., *Operational Quantum Physics*, Berlin, Springer (1995).

[9] Chiu, C. B., Misra, B., and Sudarshan, E. C. G., Time evolution of unstable quantum states and a resolution of Zeno's paradox, *Physical Review* **D16**: 520–529 (1977).

[10] de Muynck, W. M., Stoffels, W. W., and Martens, H., Joint measurement of interference and path observables in optics and neutron interferometry, *Physica B* **175**: 127–132 (1991).

[11] de Muynck, W. M., De Baere, W., and Martens, H., Interpretation of quantum mechanics, joint measurement of incompatible observables, and counterfactual definiteness, *Foundations of Physics* **24**: 1598–1663 (1994).

[12] de Muynck, W. M., *Foundations of Quantum Mechanics, an Empiricist Approach*, New York, Kluwer Academic Publishers (2002).

[13] Feynman, R. P., Negative probability, in [17, pp. 235–248].

[14] Finkelstein, D. R., *Quantum Relativity*, Berlin, Springer (1996).

[15] Healey, D. M., Jr. and Schroeck, F. E., Jr., On informational completeness of covariant localization observables and Wigner coefficients, *Journal of Mathematical Physics* **36**: 453–507 (1995).

[16] Helstrom, C. W., *Quantum Detection and Estimation Theory*, New York, Academic Press (1976).

[17] Hiley, B. J. and Peat, F. D. (eds.), *Quantum Implications: Essays in Honor of David Bohm*, London, Routledge, (1987, 1991, 1994).

[18] Kraus, K., *States, Effects and Operations*, Berlin, Springer (1983).

[19] Misra, B. and Sudarshan, E. C. G., The Zeno's paradox in quantum theory, *Journal of Mathematical Physics* **18**: 756–763 (1977).

[20] Nielsen, M. A. and Caves, C. M., Reversible quantum operations and their application to teleportation, *Physical Review* **A55**: 2547–2556 (1997).

[21] Prugovečki, E., Information-theoretical aspects of quantum measurement, *International Journal of Theoretical Physics* **16**: 321–333 (1977).

[22] Royer, A., Measurement of the Wigner function, *Physical Review Letters* **55**: 2745–2748 (1985).

[23] Royer, A., Measurement of quantum states and the Wigner function, *Foundations of Physics* **19**: 3–30 (1989).

[24] Scully, M. O., Walther, H., and Schleich, W., Feynman's approach to negative probability in quantum mechanics, *Physical Review* **A49** (1994): 1562–1566.

[25] Wheeler, J. A., The 'past' and the 'delayed-choice' double-slit experiment, in Marlow, A. R. (ed.), *Mathematical Foundations of Quantum Theory*, New York, Academic Press, 9–48 (1978).

[26] Zurek, W. H., Quantum origin of quantum jumps: breaking of unitary symmetry induced by information transfer in the transition from quantum to classical, *Physical Review* **A76**: 052110-1–5 (2007).

[27] Zwolak, M. and Zurek, W. H., Complementarity of quantum discord and classically accessible information, *Scientific Reports* **3** (2013): 1729.

5

Determination of the stationary basis from protective measurement on a single system

LAJOS DIÓSI

We generalize protective measurement for protective joint measurement of several observables. The merit of joint protective measurement is the determination of the eigenstates of an unknown Hamiltonian rather than the determination of features of an unknown quantum state. As an example, we precisely determine the two eigenstates of an unknown Hamiltonian by a single joint protective measurement of the three Pauli matrices on a qubit state.

5.1 Introduction

Protective measurement is one of the unexpected consequences of the strange structure of quantum mechanics. According to general wisdom, we cannot gain information on the unknown state $\hat{\rho}$ of a single quantum system unless we distort the state itself. In particular, we cannot learn the unknown state of a single system whatever test we apply to it. It came as a surprise that in weak measurements [1] the expectation value $\langle \hat{A} \rangle$ of an observable $\hat{A}$ can be tested on a large ensemble of identically prepared unknown states in such a way that the distortions per single systems stay arbitrarily small (see [2], too). An indirectly related surprise came with the so-called protective measurements [3, 4, 5] capable of testing $\langle \hat{A} \rangle$ at least in an unknown eigenstate of the Hamiltonian $\hat{H}$ at arbitrarily small distortion of the state itself. Interesting debates followed the proposal as to the merit of protective measurement in the interpretation of the wave function of a single system instead of a statistical ensemble (see, e.g., [6] and references therein).

My work investigates an alternative merit of protective measurement. First I construct joint protective measurements of several observables $\hat{A}_1, \hat{A}_2, \ldots$ and re-state the original equations for them in a general form. Then I show that the straightforward task that a single joint protective measurement solves on a single system is the determination of the eigenstates of an otherwise unknown Hamiltonian.

5.2 Joint protective measurement of several observables

Consider a single quantum system in state $\hat{\rho}$, and suppose that its Hamiltonian has discrete non-degenerate spectrum $\omega_1, \omega_2, \ldots$:

$$\hat{H} = \sum_n \omega_n |n\rangle \langle n|, \tag{5.1}$$

with eigenstates $|n\rangle$. Consider a set of Hermitian observables $\hat{A}_1, \hat{A}_2, \ldots$ For later convenience, introduce their expectation values in the stationary eigenstates:

$$\langle \hat{A}_\alpha \rangle_n = \langle n|\hat{A}_\alpha |n\rangle, \quad \alpha = 1, 2, \ldots \tag{5.2}$$

To simultaneously measure the observables $\hat{A}_1, \hat{A}_2, \ldots$, we use von Neumann detectors with the canonical variables $(\hat{x}_1, \hat{p}_1), (\hat{x}_1, \hat{p}_1), \ldots$, with vanishing Hamiltonians. We prepare the detectors in state $\hat{\rho}_D$ initially, such that the pointer variables $\hat{x}_1, \hat{x}_2, \ldots$ are of zero mean and of small dispersions $\delta x_1, \delta x_2, \ldots$, respectively. The conditions

$$0 < \delta x_\alpha \ll |\langle \hat{A}_\alpha \rangle_n - \langle \hat{A}_\alpha \rangle_m| \tag{5.3}$$

must be satisfied for as many pairs $n \neq m$ as possible for each detector $\alpha = 1, 2, \ldots$, to ensure that a maximum set of $\langle \hat{A}_\alpha \rangle_1, \langle \hat{A}_\alpha \rangle_2, \ldots$ can be distinguished by the detectors. Now we introduce the usual coupling $\hat{K} = \sum_\alpha \hat{p}_\alpha \hat{A}_\alpha / T$ between the observables and the detectors, respectively, with the factor $1/T$ where T is the duration of the protective measurement.

Let us evaluate the composite unitary dynamics in the interaction picture. The observables and the coupling become time-dependent:

$$\hat{A}_\alpha(t) = e^{it\hat{H}} \hat{A}_\alpha e^{-it\hat{H}}, \tag{5.4}$$

$$\hat{K}(t) = \frac{1}{T} \sum_\alpha \hat{p}_\alpha \hat{A}_\alpha(t). \tag{5.5}$$

The unitary transformation after time T reads

$$\hat{U}_T = \mathcal{T} \exp\left(-i \int_0^T \hat{K}(t) dt\right)$$

$$= \mathcal{T} \exp\left(-i \sum_\alpha \hat{p}_\alpha \int_0^T \hat{A}_\alpha(t) \frac{dt}{T}\right), \tag{5.6}$$

where $\mathcal{T}$ stands for time-ordering. Inserting

$$\hat{A}_\alpha(t) = \sum_{n,m} e^{i(\omega_n - \omega_m)t} |n\rangle \langle n|\hat{A}_\alpha |m\rangle \langle m|, \tag{5.7}$$

we find that the contribution of the off-diagonal elements become heavily suppressed when $T|\omega_n - \omega_m| \gg 1$ is satisfied for all $n \neq m$. The ideal protective measurement requires $T = \infty$; the corresponding unitary dynamics contains the contribution of diagonal elements (5.2) only:

$$\hat{U}_\infty = \sum_n \exp\left(-i \sum_\alpha \hat{p}_\alpha \langle \hat{A}_\alpha \rangle_n\right) |n\rangle \langle n|. \tag{5.8}$$

Observe that the eigenvalues ω_n of the Hamiltonian play no role, only the eigenstates $|n\rangle$ do. The dynamics of joint protective measurement of (a finite number of) observables $\hat{A}_1, \hat{A}_2, \ldots$ is captured by $\hat{U}_\infty$. It will entangle the system with the detectors in such a way that the pointer variables $\hat{x}_1, \hat{x}_2, \ldots$ get shifted by the expectation values of $\langle \hat{A}_1 \rangle_n, \langle \hat{A}_2 \rangle_n, \ldots$, taken in each eigenstate $|n\rangle$ in turn. The readout of the detectors will obtain the outcomes

$$x_1 = \langle \hat{A}_1 \rangle_n \pm \delta x_1, \quad x_2 = \langle \hat{A}_2 \rangle_n \pm \delta x_2, \quad \ldots \tag{5.9}$$

with probability $|\langle n|\psi \rangle|^2$. The terms $\pm \delta x_1, \pm \delta x_2, \ldots$ indicate statistical errors. The above outcomes mean that we have occasionally (i.e., whenever the thresholds (5.3) disclose the ambiguity of n) collapsed the state $\hat{\rho}$ of the system into $|n\rangle \langle n|$ and we have precisely (i.e., at arbitrary small errors) measured the expectation values of $\hat{A}_1, \hat{A}_2, \ldots$ in the stationary state $|n\rangle$ of $\hat{H}$.

Let us test the above dynamics on the uncorrelated pure initial state $|\psi_D\rangle |\psi\rangle$ of the system+detectors compound:

$$|\psi_D\rangle |\psi\rangle \longrightarrow \hat{U}_\infty |\psi_D\rangle |\psi\rangle. \tag{5.10}$$

Let us introduce the wave function $\psi_D(x_1, x_2, \ldots)$ of the detectors. If we substitute (5.8) and multiply both sides of (5.10) by $\langle x_1, x_2, \ldots |$, we get

$$\psi_D(x_1, x_2, \ldots) |\psi\rangle \longrightarrow \sum_n \psi_D(x_1 - \langle \hat{A}_1 \rangle_n, x_2 - \langle \hat{A}_2 \rangle_n, \ldots) |n\rangle \langle n|\psi\rangle.$$

This shows that, under the conditions (5.3) on the initial wave function $\psi_D(x_1, x_2, \ldots)$ of the detectors, the state on the r.h.s. prepares the von Neumann measurement of n and $\langle \hat{A}_1 \rangle_n, \langle \hat{A}_2 \rangle_n, \ldots$ In particular, the initial probability density $P(x_1, x_2, \ldots) = |\psi_D(x_1, x_2, \ldots)|^2$ changes like this:

$$P(x_1, x_2, \ldots) \longrightarrow \sum_n |\langle n|\psi\rangle|^2 P(x_1 - \langle \hat{A}_1 \rangle_n, x_2 - \langle \hat{A}_2 \rangle_n, \ldots).$$

Formally, this expression is the statistical mixture corresponding to a von Neumann projective measurement of the stationary basis resulting in the outcome n with probability $|\langle n|\psi\rangle|^2$. In each term the initial positions of the pointers get shifted by the expectation values of the corresponding observables in the given post-measurement eigenstate $|n\rangle$. The eigenvalues ω_n themselves do not appear in the result, since they already canceled from the unitary dynamics $\hat{U}_\infty$, as we observed before.

5.3 Protective measurement of the stationary basis

We are going to show that a single joint protective measurement determines the eigenstates of the unknown Hamiltonian. Our example is a single qubit in an unknown initial state

$$\hat{\rho} = \frac{1}{2}(\hat{I} + \vec{s}\,\hat{\vec{\sigma}}), \quad |\vec{s}| \leq 1, \tag{5.11}$$

with unknown spatial polarization vector $\vec{s}$. Unknown is the Hamiltonian as well:

$$\hat{H} = \Omega\vec{e}\,\hat{\vec{\sigma}}, \quad |\vec{e}| = 1, \tag{5.12}$$

with unknown strength Ω and unknown direction $\vec{e}$ of the external "magnetic" field. The Hamiltonian has two unknown eigenvalues $\pm\Omega$ and eigenstates $|\pm\rangle$:

$$\hat{H} = \Omega|+\rangle\,\langle+| - \Omega|-\rangle\,\langle-| \equiv \Omega\frac{\hat{I} + \vec{e}\,\hat{\vec{\sigma}}}{2} - \Omega\frac{\hat{I} - \vec{e}\,\hat{\vec{\sigma}}}{2}. \tag{5.13}$$

Now we prepare three von Neumann detectors and couple them to the three qubit observables $\hat{A}_\alpha = \hat{\sigma}_\alpha$, for $\alpha = 1, 2, 3$, respectively. Their joint protective measurement is described by the unitary operator (5.8):

$$\hat{U}_\infty = \exp\left(-i\hat{\vec{p}}\langle\hat{\vec{\sigma}}\rangle_+\right)|+\rangle\,\langle+| + \exp\left(-i\hat{\vec{p}}\langle\hat{\vec{\sigma}}\rangle_-\right)|-\rangle\,\langle-|. \tag{5.14}$$

With $\langle\hat{\vec{\sigma}}\rangle_\pm = \langle+|\hat{\vec{\sigma}}|+\rangle = \pm\vec{e}$, the coupling shows the following simple dependence on the unknown parameter $\vec{e}$ of $\hat{H}$:

$$\hat{U}_\infty = \exp\left(-i\hat{\vec{p}}\,\vec{e}\right)|+\rangle\,\langle+| + \exp\left(+i\hat{\vec{p}}\,\vec{e}\right)|-\rangle\,\langle-|. \tag{5.15}$$

This unitary operator acts on the initial uncorrelated state:

$$\hat{\rho}_D\hat{\rho} \longrightarrow \hat{U}_\infty\hat{\rho}_D\hat{\rho}\hat{U}_\infty^\dagger. \tag{5.16}$$

Let the state $\hat{\rho}_D$ be constrained by $\delta x_1, \delta x_2, \delta x_3 \ll 1$; see (5.3). Inserting (5.15) and taking the diagonal matrix element $\langle\vec{x}|\ldots|\vec{x}\rangle$ of both sides, we get the resulting change of the initial pointer statistics $P(\vec{x}) = \langle x|\hat{\rho}_D|x\rangle$:

$$P(\vec{x}) \longrightarrow |\langle+|\hat{\rho}|+\rangle|^2 P(\vec{x} - \vec{e}) + |\langle-|\hat{\rho}|-\rangle|^2 P(\vec{x} + \vec{e}). \tag{5.17}$$

Expressing $|\langle\pm|\hat{\rho}|\pm\rangle|^2$ via (5.11) and (5.13), the final statistics of the pointers x_1, x_2, x_3 becomes

$$\frac{1 + \vec{e}\,\vec{s}}{2}P(\vec{x} - \vec{e}) + \frac{1 - \vec{e}\,\vec{s}}{2}P(\vec{x} + \vec{e}). \tag{5.18}$$

If we read out the three detectors, the outcome is $\vec{x} \approx \pm\vec{e}$ with probability $(1\pm\vec{e}\,\vec{s})/2$, respectively. We have thus determined the spatial direction $\vec{e}$ of the external field at arbitrary high precision up to its sign, however. The precision of the measured components e_1, e_2, e_3 is given respectively by the initial dispersions $\delta x_1, \delta x_2, \delta x_3 \ll 1$;

it does not depend on the initial state $\hat{\rho}$ of the qubit. The strength Ω of the field remains unknown while the obtained knowledge of $\pm\vec{e}$ means that we have precisely inferred the two stationary states $|\pm\rangle$. Our protective measurement collapses the system, exactly like an ideal von Neumann measurement of $\hat{H}$ would do, into one of the two stationary states, but we cannot learn into which one of the two.

5.4 Summary

I have generalized the concept of protective measurement for joint protective measurement of a (possibly finite) number of observables, determined the corresponding unitary operation and its action on the arbitrary uncorrelated initial state of the system and the detectors. I have shown that on a single qubit of unknown state and unknown Hamiltonian, the two stationary states can be determined in a single joint protective measurement of the three Pauli matrices. The post-measurement state of the qubit is just like it would be after a projective measurement of $\hat{H}$. This result may certainly be generalized for higher dimensional systems as well. In fact, the full Hamiltonian can always be determined on a single system if, e.g., we perform a suitable sequence of standard measurements. Yet the surprising feature of the joint protective measurement is that the stationary states can be determined in a single step and in a transparent model.

This work was supported by the Hungarian Scientific Research Fund under Grant No. 75129 and the EU COST Action MP1006.

References

[1] Aharonov, Y., Albert, D. Z., and Vaidman, L. (1988) How the result of measurement of a component of the spin of a spin-1/2 particle can turn out to be 100. *Physical Review Letters* **60**, 1351–1354.

[2] Diósi, L. (2006) Quantum mechanics: weak measurements. *Encyclopedia of Mathematical Physics*, eds. J.-P. Françoise, G. L. Naber, and S. T. Tsou, Oxford: Elsevier, vol. 4, 276–282.

[3] Aharonov, Y. and Vaidman, L. (1993) Measurement of the Schrödinger wave of a single particle. *Physics Letters A* **178**, 38–42.

[4] Aharonov, Y., Anandan, J., and Vaidman, L. (1993) Meaning of the wave function. *Physical Review A* **47**, 4616–4626.

[5] Aharonov, Y., Anandan, J., and Vaidman, L. (1996) The meaning of protective measurements, *Foundations of Physics* **26**, 117–126.

[6] Gao, S. (2013) Protective measurement: a paradigm shift in understanding quantum mechanics. *url = philsci-archive.pitt.edu/9627/*.

6

Weak measurement, the energy–momentum tensor and the Bohm approach

ROBERT FLACK AND BASIL J. HILEY

In this chapter we show how the weak values, $\langle \boldsymbol{x}(t)|\widehat{P^\mu}|\psi(t_0)\rangle / \langle \boldsymbol{x}(t)|\psi(t_0)\rangle$, are related to the $T^{0\mu}(\boldsymbol{x}, t)$ component of the energy–momentum tensor. This enables the local energy and momentum to be measured using weak measurement techniques. We also show how the Bohm energy and momentum are related to $T^{0\mu}(\boldsymbol{x}, t)$ and therefore it follows that these quantities can also be measured using the same methods. Thus the Bohm "trajectories" can be empirically determined, as was shown by Kocsis *et al.* (2011a) in the case of photons. Because of the difficulties with the notion of a photon trajectory, we argue the case for determining experimentally similar trajectories for atoms where a trajectory does not cause these particular difficulties.

6.1 Introduction

The notion of weak measurement introduced by Aharonov, Albert and Vaidman (1988) and Aharonov and Vaidman (1990) has opened up a radically new way of exploring quantum phenomena. In contrast to the strong measurement (von Neumann, 1955), which involves the collapse of the wave function, a weak measurement induces a more subtle phase change which does not involve any collapse. This phase change can then be amplified and revealed in a subsequent strong measurement of a complementary operator that does not commute with the operator being measured. This amplification explains why it is possible for the result of a weak spin measurement of a spin-1/2 atom to be magnified by a factor of 100 (Aharonov *et al.*, 1988; Duck, Stevenson and Sudarshan, 1989). A weak measurement, then, provides a means of amplifying small signals as well as allowing us to gain new, more subtle information about quantum systems.

One of the new features that we will concentrate on in this chapter is the possible measurement of the $T^{0\mu}(\boldsymbol{x}, t)$ components of the energy–momentum tensor.

In Section 6.3.2 we show that these components are related to the real part of the weak value $\langle P^\mu(t)\rangle_W$ through the expression

$$\mathfrak{R}\langle P^\mu(t)\rangle_W = T^{0\mu}(\boldsymbol{x}, t), \tag{6.1}$$

where the weak value is defined by

$$\langle P^\mu(t)\rangle_W := \frac{\langle \boldsymbol{x}(t)|\widehat{P^\mu}|\psi(t_0)\rangle}{\langle \boldsymbol{x}(t)|\psi(t_0)\rangle}.$$

This allows us not only to discuss, but also to actually measure, the *local* properties of the energy, $E(\boldsymbol{x}, t)$, and the momentum, $P(\boldsymbol{x}, t)$. These results are in contrast with the standard treatments, which only discuss the *global* properties, using the expressions

$$E(t) = \int T^{00}(\boldsymbol{x}, t)\mathrm{d}^3 x \qquad \text{and} \qquad P^j(t) = \int T^{0j}(\boldsymbol{x}, t)\mathrm{d}^3 x.$$

The use of these global quantities is usually justified by claiming that in quantum mechanics, it is only through these global quantities that energy and momentum can be uniquely defined and conserved (Schweber, 1961).

However, we can show that the *local* energy and momentum can also be conserved. The way to do this comes from a surprising direction – the Bohm approach (Bohm, 1952a, 1952b). Critically examining the mathematical structure in the simple case of the non-relativistic, spin-zero particle, we find the real part of the Schrödinger equation under polar decomposition gives the quantum Hamilton–Jacobi equation (qHJ),[1] namely,

$$\frac{\partial S}{\partial t} + \frac{(\nabla S)^2}{2m} + Q + V = 0, \tag{6.2}$$

where Q is a novel form of energy, which we call the quantum potential energy. This takes the form

$$Q(\boldsymbol{x}, t) = -\frac{\nabla^2 R(\boldsymbol{x}, t)}{2mR(\boldsymbol{x}, t)}.$$

If we now follow Bohm and define

$$P_{\mathrm{B}}(\boldsymbol{x}, t) := \nabla S(\boldsymbol{x}, t) \qquad \text{and} \qquad E_{\mathrm{B}}(\boldsymbol{x}, t) := -\partial_t S(\boldsymbol{x}, t), \tag{6.3}$$

we then find that the qHJ equation becomes a simple local energy conservation equation

$$E_{\mathrm{B}}(\boldsymbol{x}, t) = (P_{\mathrm{B}}(\boldsymbol{x}, t))^2/2m + Q(\boldsymbol{x}, t) + V(\boldsymbol{x}, t),$$

---

[1] We use $\hbar = 1$ throughout this chapter.

provided we regard the quantum potential energy as a new quality of energy appearing only in the quantum domain. Indeed, when $Q = 0$, we have the classical equation for the conservation of energy. In this case S becomes the classical action and we recapture the classical equations of motion.

Hiley and Callaghan (2010a, 2010b) have shown further that Bohm's conjectured expressions for the momentum and energy given by (6.3) can be put on a firmer footing by showing that

$$\rho P_B^j(x, t) = T^{0j}(x, t) \qquad \text{and} \qquad \rho E_B(x, t) = T^{00}(x, t), \tag{6.4}$$

where ρ is the probability density and where expressions for P_B^j and E_B are obtained from the expressions of the Lagrangian for the Pauli and Dirac particles respectively. A more detailed explanation of these results will be given in Section 6.3.

Thus our results hold not only for Schrödinger particles, but also for Pauli and Dirac particles. The corresponding real parts of the Pauli and Dirac equations give rise to their respective qHJ equations. Both these equations contain their corresponding expressions for the quantum potential energies (see Hiley and Calaghan, 2012). Thus the quantum potential energy, the existence of which was regarded as *ad hoc* by Heisenberg (1958) and unnecessary by Dürr *et al* (1996), plays a vital role in the conservation of local energy and is at the heart of quantum theory.

A comparison of Equations (6.1) and (6.4) shows the relation between the weak values and the parameters introduced by Bohm; for example, the weak value $\langle P_x \rangle_W$ is exactly the x-component of the Bohm momentum. Indeed, Leavens (2005), Wiseman (2007) and Hiley (2012) have already shown that this is a particular example of a more general result applicable to the Schrödinger particle so that analogous weak values can be found for the corresponding Bohm energy and Bohm kinetic energy. Similar relations also apply to relativistic particles with spin (Hiley and Callaghan, 2012). This shows that these quantities are not arbitrarily added "philosophical" terms, but actually correspond to entities that can be measured in the laboratory.

Thus the Bohm approach, introduced originally to show that it is possible to provide a realistic model of quantum phenomena without the need for the observer to play an essential role in the theory, can no longer be criticized on the grounds that it uses unobservable terms like the Bohm momentum and Bohm energy. In some cases these quantities have been measured using weak measurement techniques (Kocsis *et al.*, 2011a). Furthermore, the criticism that the notion of particles following unobserved "trajectories" adds no new physical content to the theory can also no longer be sustained. Thus what appeared to be an empty physical theory actually adds new insight to standard quantum mechanics, and is not in opposition to it as is often perceived. It actually enriches the standard theory.

The recent experimental results of Kocsis *et al* (2011a) confirm that the usual criticism can no longer hold. These authors empirically determine an ensemble of what they call "photon trajectories" in regions of interference where standard arguments using the uncertainty principle would suggest that the interference must be destroyed (see Figure 6.1).

Although the notion of a trajectory for a photon is not without its own difficulties, these photon "trajectories" have a striking resemblance to the particle trajectories calculated in Philippidis, Dewdney and Hiley (1979) for a Schrödinger particle in the Bohm model (see Figure 6.2). The question we want to discuss in this chapter is whether we can use these techniques for atoms rather than photons. In this case

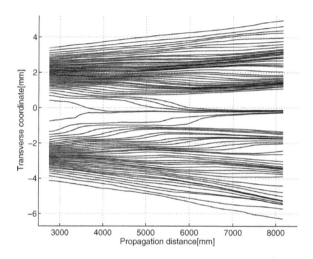

Figure 6.1 Experimentally produced photon "trajectories" (see Kocsis *et al.*, 2011a).

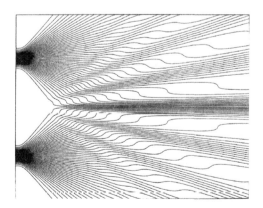

Figure 6.2 Theoretical Schrödinger particle trajectories (see Philippidis *et al.*, 1979).

difficulties associated with photon trajectories do not arise. Using atoms in the non-relativistic region will enable us to make a direct comparison with the theoretical calculations of Philippides *et al.* (1979). We discuss our experimental proposals in Section 6.5.

6.2 Quantum measurement

6.2.1 von Neumann measurement

The usual notion of a quantum measurement, the von Neumann measurement, involves a process in which the wave function "collapses" into one of the eigen-functions of the operator whose value is required. Let us recall how this works in some detail to enable us to directly contrast it with what is involved in a weak measurement.

Suppose we want to measure a property of our system which is described by an operator $\hat{A}$ such as, for example, the spin of a particle. One way to determine this is to pass the particle through an inhomogeneous Stern–Gerlach magnetic field which deflects the particles according to their spin state. If the particles are initially traveling along the y-axis, then we orient the field in a transverse direction relative to the line of flight, say, along the z-axis. The field then separates the particles into their two spin states. The "pointer reading" then corresponds to the position of the final two peaks, spin up and spin down.

With the Stern–Gerlach experiment (SG) in mind, we follow Bohm (1951) and use the interaction Hamiltonian $H_1 = g(t)\hat{Z}\,\hat{\sigma}_z$, so that the position operator $\hat{Z}$ is coupled to the spin operator $\hat{\sigma}_z$. $g(t)$ is some function describing the strength of the interaction. The problem then is to solve the Schrödinger equation using this Hamiltonian. We remind readers of how this is done in a standard (strong) measurement simply to contrast it to a weak measurement.

We introduce the time development operator, $U(t, t_0)$, to determine the final ket $|\Psi(t)\rangle$ from some initial ket $|\Psi(t_0)\rangle$, so that

$$|\Psi(t)\rangle = U(t, t_0)|\Psi(t_0)\rangle.$$

$U(t, t_0)$ satisfies

$$\frac{\partial U(t, t_0)}{\partial t} = -\mathrm{i}H(t)U(t, t_0)$$

which gives the solution

$$U(t, t_0) = \exp\left[-\mathrm{i}\int_{t_0}^{t} H(t')\mathrm{d}t'\right].$$

We now substitute the interaction Hamiltonian with the condition

$$\eta = \int_{t_0}^{t} g(t')dt' = D\Delta t,$$

where D is a measure of the strength of the interaction and Δt is the time the interaction is active. This gives the final state as

$$|\Psi(t)\rangle = \exp\left[-i\eta\hat{Z}\,\hat{\sigma}_z\right]|\Psi(t_0)\rangle. \tag{6.5}$$

Then

$$\langle z|\Psi(t)\rangle = \int \langle z|\exp\left[-i\eta\hat{Z}\,\hat{\sigma}_z\right]|z'\rangle\langle z'|\Psi(t_0)\rangle dz',$$

so that we have

$$\Psi(z,t) = \exp\left[-i\eta z\hat{\sigma}_z\right]\Psi(z,t_0).$$

Let us choose the total initial spin state to be

$$\Psi(z,t_0) = \psi(z,t_0)\sum_{n=1}^{2} c_n|\xi_n\rangle,$$

where $|\xi_n\rangle$ are the eigenkets of $\hat{\sigma}_z$, viz. $\hat{\sigma}_z|\xi_n\rangle = a_n|\xi_n\rangle$ where, for a spin-half system, a_n takes the value $+1$ or -1. The final state will be

$$\Psi(z,t) = c_+\psi(z-\eta) + c_-\psi(z+\eta). \tag{6.6}$$

Thus we are left with two sharply peaked wave packets, one centered at $\eta = D\Delta t$ and the other at $\eta = -D\Delta t$. In the case when the initial wave function is Gaussian, $|\psi(t_0)\rangle = \int \exp\left[-\frac{z'^2}{4(\Delta z)^2}\right]|z'\rangle dz'$, the final wave function in the momentum representation would take the form

$$\phi(p,t) = A_+ \exp[-(\Delta z)^2(p^2 - \eta)] + A_- \exp[-(\Delta z)^2(p^2 + \eta)]. \tag{6.7}$$

If the interaction is sufficiently strong to ensure the two wave packets are well separated, the "pointer" position can distinguish the two spin states (see Figure 6.3).

Each particle will end up in one of the wave packets and will be detected at the appropriate point on the screen. This means that at each detection, the wave function "collapses" into one or other of the two packets described in Equation (6.6), in which it then remains. It is this process that is called a von Neumann or strong measurement.

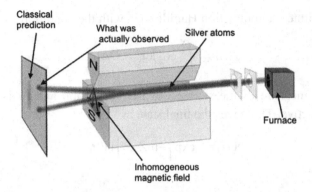

Figure 6.3 Diagram of the classic Stern–Gerlach experiment. A beam of silver atoms is produced in a furnace and passed through a very strong inhomogeneous magnetic field. The beam is split into its two spin components.

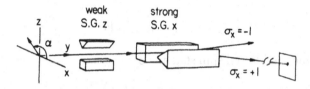

Figure 6.4 The modified SG apparatus to observe and measure the weak value for spin-1/2 particles. The diagram clearly shows the three stages of the weak measurement process; preparation, weak and strong (Duck *et al.*, 1989).

6.2.2 *Example of weak measurements for spin*

In order to make a weak measurement, we must reduce the strength of the field of the Stern–Gerlach magnet to the point where the two wave packets do not completely separate, but still overlap so that interference effects will be present. More precisely, if the incident wave packet is a Gaussian, the spatial part of the packets become separated in the z-direction. If the separation of their centers is, say, a distance a, and this is less than the width, Δz, of the incoming wave packet, the packets will interfere. Detailed calculations sketched below show that the resulting beam approximates to a Gaussian, but with its center displaced by an amount that depends on the weak value $\langle \sigma_z \rangle_{\mathrm{W}} = \langle \xi_{\mathrm{f}} | \hat{\sigma}_z | \xi_{\mathrm{i}} \rangle / \langle \xi_{\mathrm{f}} | \xi_{\mathrm{i}} \rangle$.

Further interesting results can be obtained if we introduce a further Stern–Gerlach magnet to make a strong measurement in, say, the x-direction. In this case the beam is split into two, this time along the x-direction. If these two beams are allowed to separate completely, then we see two Gaussians whose centers are displaced from two Gaussians that would have been produced if the weak Stern–Gerlach magnet had not been in place. The displacements of each packet enable the weak value $\langle \sigma_z \rangle_{\mathrm{W}} = \langle \xi_{\mathrm{f}} | \hat{\sigma}_z | \xi_{\mathrm{i}} \rangle / \langle \xi_{\mathrm{f}} | \xi_{\mathrm{i}} \rangle$ to be measured.

Figure 6.4 illustrates how all of this comes about. A beam of neutral atoms, moving along the y-axis, produced from an oven, are first polarized at an angle α to the x-axis. The beam then passes through an inhomogeneous magnetic field oriented along the z-axis, SGz, which produces a small separation along the z-axis. By small here we mean the center of each packet is separated by less than the width of each packet.

Finally, the weakly separated packets enter another magnet, SGx, this time with its field oriented along the x-axis. Here a strong measurement takes place along the x-axis. Note that SGz and SGx are measuring conjugate variables.

6.2.3 Details of the weak measurement of spin

Having outlined what the outcome of a weak measurement of spin will give us, let us now look into the details. We start, again, with the interaction Hamiltonian $H_I = g(t)\hat{Z}\hat{\sigma}_z$, which, as we have seen, gives a final state

$$|\psi(t)\rangle|\xi_f\rangle = e^{i\eta\hat{Z}\hat{\sigma}_z}|\psi(t_0)\rangle|\xi_i\rangle.$$

Now multiplying through by $\langle\xi_f|$ we find

$$|\psi(t)\rangle = \langle\xi_f|e^{i\eta\hat{Z}\hat{\sigma}_z}|\xi_i\rangle|\psi(t_0)\rangle.$$

Expanding the expontional in the first term on the r.h.s. of this equation, we find

$$\langle\xi_f|e^{i\eta\hat{Z}\hat{\sigma}_z}|\xi_i\rangle = \langle\xi_f|\xi_i\rangle \sum_{m=0}^{\infty} \frac{(i\eta\hat{Z})^m}{m!} \frac{\langle\xi_f|\hat{\sigma}_z^m|\xi_i\rangle}{\langle\xi_f|\xi_i\rangle}.$$

We can now rewrite this expression as

$$\langle\xi_f|e^{i\eta\hat{Z}\hat{\sigma}_z}|\xi_i\rangle = \langle\xi_f|\xi_i\rangle \left[e^{i\eta\hat{Z}\langle\sigma_z\rangle_W} + \sum_{m=2}^{\infty} \frac{(i\eta\hat{Z})^m}{m!} \left[\langle(\sigma_z)^m\rangle_W - \langle\sigma_z\rangle_W^m \right] \right], \quad (6.8)$$

where $\langle(\sigma_z)^m\rangle_W$ is the "weak value" of the operator $(\hat{\sigma}_z)^m$ (see Aharonov *et al.*, 1988). When the term involving the sum $\sum_{m=2}$ in Equation (6.8) is small and can be neglected, we can write the final state at the position z in the form

$$\langle\xi_f|e^{i\eta\hat{Z}\hat{\sigma}_z}|\xi_i\rangle = \langle\xi_f|\xi_i\rangle e^{i\eta\hat{Z}\langle\sigma_z\rangle_W}.$$

If we choose the initial position wave function to be a Gaussian so that

$$|\psi(t_0)\rangle = \int \exp\left[-\frac{z'^2}{4(\Delta z)^2}\right]|z'\rangle dz'$$

in the z-representation, then we have

$$\psi(z, t) = e^{i\eta z\langle\sigma_z\rangle_W} \exp\left[-\frac{z^2}{4(\Delta z)^2}\right].$$

Recall that the weak value can be a complex number $\langle\sigma_z\rangle_W = \mathfrak{R}\langle\sigma_z\rangle_W + i\mathfrak{I}\langle\sigma_z\rangle_W$. This means that the imaginary part of the weak value shifts the real part of the wave function to be centered around $z = 2(\Delta z)^2\mathfrak{I}\langle\sigma_z\rangle_W$ since, in this case, the wave function is

$$\psi(z,t) \propto \exp\left[-\frac{(z + 2(\Delta z)^2\eta\mathfrak{I}\langle\sigma_z\rangle_W)^2}{4(\Delta z)^2}\right].$$

If we want to evaluate the real part of the weak value, we need to consider a wave function in the p-representation when we find

$$\phi(p,t) \propto \exp[-(\Delta z)^2(p - \eta\mathfrak{R}\langle\sigma_z\rangle_W)^2]. \tag{6.9}$$

In order to analyze the results further we must pass the beam through a Stern–Gerlach magnet aligned in the x-direction. This will split the beam into two components which will enable us to measure either

$$\langle\sigma_z\rangle_{W(+x)} = \frac{\langle +x|\hat{\sigma}_z|\xi_i\rangle}{\langle +x|\xi_i\rangle} \quad\text{or}\quad \langle\sigma_z\rangle_{W(-x)} = \frac{\langle -x|\hat{\sigma}_z|\xi_i\rangle}{\langle -x|\xi_i\rangle}.$$

If the final screens shown in Figure 6.4 are placed sufficiently far from the last magnet (strong SGx), so that the displacement in the x-direction will be greater than width of the wave packet, we will be able to measure the displacement of each beam from which we can determine the respective weak values.

Let us now compare the result given by Equation (6.9) with the first term of Equation (6.7). We see that the centroid of the Gaussian is shifted by an amount that depends upon the initial angle of polarisation α. Duck *et al.* (1989) show that by a specific choice of α, the difference can be greater by a factor of 70.

What we see from the above results is that although the spin part of the quantum state has collapsed, the spatial part does not collapse, it is shifted as a whole. The shift of the maximum gives us the weak value, $\langle\sigma_z\rangle_W$. This is a clear example of what Aharonov and Vaidman (1993, 1995) call a "protected" wave function.

Of course, in actually carrying out the experiment we have to ensure that the neglected terms in Equation (6.8) are satisfied. Also, care must be taken to ensure that the terms $\partial B_x/\partial x$ and $\partial B_y/\partial y$, which cannot vanish because of the condition $\nabla \cdot B = 0$, are taken into account. Detailed discussions of these effects is presented in Aharonov, Albert and Vaidman (1988) and in Duck, Stevenson and Sudarshan (1989) and will not be discussed further here.

If all these conditions are satisfied, the theory predicts that the x-displacement is greater than expected, producing an amplification effect. If the separation of the spin-z-up and spin-z-down wave packets produced in a normal Stern–Gerlach is d, then the weak measurement process should separate them by up to a distance $70d$.

We see here how the amplification generated by the weak measurement comes about. Notice that no external amplification process is being used, only a subtle manipulation of the wave function of the atom occurs. In a sense this could be viewed as *self-amplification* with the advantage that no noise is generated, unlike real-life amplifiers where irreducible distortion of the signal is inevitable. As we go deeper and deeper into the atomic and sub-atomic world, signals are getting smaller and smaller. Therefore the weak measurement technique could offer the possibility of a new way of making observations that are hidden by insufficient resolution and the noise of the measuring instruments.

6.2.4 Experimental realization of weak Stern–Gerlach measurement using photons

The first experiments to test these ideas have used optical analogues. For example Ritchie, Story and Hulet (1991) used a Gaussian-mode laser source and the Stern–Gerlach magnets were replaced by optical polarizers. The weak measurement was performed by a thin birefringent-crystalline quartz plate as shown in Figure 6.5. The analysis produces a new feature which must be taken into account.

The laser light traveling, in this case, along the z-axis is polarized at an angle α to the x-axis. The electric field of the input light is

$$E_i = E_0 \exp\left[-\frac{x^2 + y^2}{\Delta^2}\right](\cos \alpha \hat{x} + \sin \alpha \hat{y}),$$

where Δ is the width of the beam. The birefingent plate is a plane-parallel uniaxial crystal whose optic axis is along the x-axis, so that the ordinary ray goes straight through, while the extraordinary ray is deflected in the y-direction. As a result the two components corresponding to different polarization states become separated

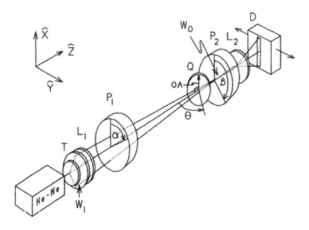

Figure 6.5 Weak Stern–Gerlach measurement using photons (Ritchie *et al.* 1991). P_1 and P_2 are the polarisers, OA is the thin uniaxial birefringent crystal responsible for the weak effect.

by a distance a, which is made small compared with the overall beam width Δ. The emerging beam is then described by

$$E_W = E_0 \exp\left[\frac{-x^2}{\Delta^2}\right]\left[\cos\alpha \exp\left[\frac{-(y+a)^2}{\Delta}\right]e^{i\phi}\hat{x} + \sin\alpha \exp\left[\frac{-y^2}{\Delta^2}\right]\hat{y}\right].$$

Here ϕ is the phase difference arising from the difference in the optical path lengths of the two rays.

The post selection is made using a polarizer aligned at an angle β to the x-axis. The resulting wave function is

$$E_f = E_0 \exp\left[\frac{-x^2}{\Delta^2}\right]\left[\cos\alpha\cos\beta \exp\left[\frac{-(y+a)^2}{\Delta}\right]e^{i\phi} + \sin\alpha\sin\beta \exp\left[\frac{-y^2}{\Delta^2}\right]\right]$$
$$\times(\cos\beta\hat{x} + \sin\beta\hat{y}).$$

Thus we see in this example that there are two distinct effects produced by the uniaxial crystal. First, there is a deflection in the y-direction of one component of the beam relative to the other. Second, there is a new feature, namely, a phase change between the two components due to the difference of the speeds in the two components in the crystal. In the magnetic analogue discussed earlier, only one of these effects is present, namely, a deflection between the two spin components. There is no phase factor change in the magnetic case.

If we choose $\alpha = \beta$, we find the two Gaussians add constructively to produce an intensity that is approximately an unshifted Gaussian. Suppose we rotate the second polarizer to an angle $\beta = \alpha + \pi/2 + \epsilon$ with $\epsilon \ll 1$ and suppose $a/2\Delta \ll \epsilon \ll 1$; the interference will produce a single Gaussian with its center shifted by the weak value $\langle S_y \rangle_W \approx a\cot(\epsilon)/2$, which is much larger than a. In the actual experiments an amplification of about 100 could be achieved.

To summarize, a weak measurement involves measuring a change of phase of the wave function produced by an operator rather than inducing a collapse of the wave function into an eigenfunction of that particular operator. But this does not mean that we have avoided a strong measurement altogether. The final stage of the measurement is a strong measurement; in the case we are discussing, it is a spin measurement in the x-direction. This final process collapses the *spin* part of the wave function, but leaves the form of the spatial part unchanged but displaced as a whole. It is from this displacement that the weak value can be found. Notice, to find the shape of the wave function we must perform a series of measurements on different particles, each being produced with the same initial wave function. The fact that the result is obtained statistically is not a problem. A pair of numerical values can be given to the weak values.

Having illustrated how weak measurements can be realized, we now need to explain exactly how these weak values are related to the conventional approach (see Hosoya and Shikano, 2010) and then to show how these weak values are related to the Bohm energy and momentum used in the Bohm model (see Bohm and Hiley, 1993).

6.2.5 Weak values

Let us now examine the meaning of weak values defined by

$$\langle A(t) \rangle_{\mathrm{W}} = \frac{\langle \phi(t)|\hat{A}|\psi(t_0)\rangle}{\langle \phi(t)|\psi(t_0)\rangle}$$

in more detail. From their very definition, it seems rather strange to call them "values", as they are complex numbers. They are weighted transition probability amplitudes and we can use them to give us an expression for the probability amplitude of going from $|\psi\rangle$ to $|\phi\rangle$. This latter state is called the "post-selected" state. Again the weighting function, $\langle \phi|\psi\rangle$, in the denominator is strange, since its magnitude depends on the relation between the pre- and post-selected states. This means that in some situations it can be chosen to make the weak value large by choosing the post-selected state $|\phi(t)\rangle$ to be nearly orthogonal to the initial state $|\psi(t_0)\rangle$, giving rise to the amplification discussed earlier.

To bring out the meaning of weak values more clearly, let us first note that the usual expression for the mean value of the Hermitian operator $\hat{A}$, sometimes known as a bilinear invariant, is a weighted sum of transition probability amplitudes $\langle \phi_j|\hat{A}|\psi\rangle$ and we can write

$$\langle \psi|\hat{A}|\psi\rangle = \sum_j \langle \psi|\phi_j\rangle\langle \phi_j|\hat{A}|\psi\rangle,$$

where $|\phi_j\rangle$ is a complete set of orthonormal states.

However, the mean value can also be written in the form (see Hosoya and Shikano, 2010)

$$\langle \psi|\hat{A}|\psi\rangle = \sum \langle \psi|\phi_j\rangle \left(\frac{\langle \phi_j|\psi\rangle}{\langle \phi_j|\psi\rangle} \right) \langle \phi_j|\hat{A}|\psi\rangle = \sum \rho_j(\phi) \frac{\langle \phi_j|\hat{A}|\psi\rangle}{\langle \phi_j|\psi\rangle}. \qquad (6.10)$$

Here $\rho_j(\phi)$ is the probability of finding the system in the state $|\phi_j\rangle$. Now we can see that if $|\phi_j\rangle$ and $|\psi\rangle$ are close to being orthogonal, the probability of finding a large weak value is small even though the "weak value" itself may be large. In other words, large numerical values for the weak value make a correspondingly small contribution to the overall mean value. The puzzling feature is why these weak values should be of any interest at all. Before going on to discuss this, let us make one more point.

To obtain a more comfortable feel for the weak values, notice the special case when we choose the $|\phi_j\rangle$ to be a set of eigenfunctions of the operator $\hat{A}$, i.e. $\hat{A}|a_j\rangle = a_j|a_j\rangle$, then we obtain the standard result

$$\langle\psi|\hat{A}|\psi\rangle = \sum \rho_j(a)a_j.$$

Thus the meaning of post-selection with a set of eigenstates of $\hat{A}$, namely $|a_j\rangle$, is very clear, but the meaning when a different set of states, $|\psi_j\rangle$, are post-selected is still not clear. However, certain weak values do have a specific meaning in the context of the Bohm model, as we now show, but first we must have a closer look at the meaning of bilinear invariants from a more general point of view.

6.3 Bilinear invariants

6.3.1 Bilinear invariants of the second kind

So far we have confined our discussion to bilinear invariants of the form $\langle\psi|\hat{A}|\psi\rangle$. However, Takabayasi (1955) points out that when the operators $\hat{A}$ are represented as derivatives, we need to introduce what he calls "bilinear invariants of the second kind" in order to have a complete specification of a quantum system. These invariants can be written in the form

$$(\partial\psi)\psi^* \pm \psi(\partial\psi^*).$$

Here we have written a generic ∂ for operators like $\partial/\partial x$, $\partial^2/\partial x^2$, $\partial/\partial t, \ldots$ If we take the plus sign, we will simply have a derivative, which will give us a bilinear invariant of the first kind. Taking the minus sign gives us a bilinear invariant of the second kind. We will follow convention and write this as

$$\psi \overset{\leftrightarrow}{\partial} \psi^* = (\partial\psi)\psi^* - \psi(\partial\psi^*).$$

To understand the meaning of these invariants we need to return to consider the energy–momentum tensor.

6.3.2 The energy–momentum tensor

As is well known, the Schrödinger equation can be derived from the Lagrangian

$$\mathcal{L} = -\frac{1}{2m}\nabla\psi^* \cdot \nabla\psi + \frac{i}{2}[(\partial_t\psi)\psi^* - (\partial_t\psi^*)\psi] - V\psi^*\psi.$$

The details will be found in Heisenberg (1949). Using the energy–momentum tensor defined by

$$T^{\mu\nu} = -\left\{\frac{\partial\mathcal{L}}{\partial(\partial^\mu\psi)}\partial^\nu\psi + \frac{\partial\mathcal{L}}{\partial(\partial^\mu\psi^*)}\partial^\nu\psi^*\right\},$$

we find the momentum density can be written as

$$T^{0j} = -\left\{\frac{\partial\mathcal{L}}{\partial(\partial^0\psi)}\partial^j\psi + \frac{\partial\mathcal{L}}{\partial(\partial^0\psi^*)}\partial^j\psi^*\right\}.$$

Because

$$\frac{\partial \mathcal{L}}{\partial(\partial^0 \psi)} = \frac{1}{2\mathrm{i}} \psi^* \quad \text{and} \quad \frac{\partial \mathcal{L}}{\partial(\partial^0 \psi^*)} = -\frac{1}{2\mathrm{i}} \psi,$$

we find

$$T^{0j} = \frac{\mathrm{i}}{2} \left[\psi^* \partial^j \psi - \psi \partial^j \psi^* \right].$$

We see immediately that this is the expression for the current, as expected. If we write $\partial^j = -\nabla$, we immediately recognise that $T^{0j}(x, t)$ is the Bohm momentum, $p_B(x, t) = \nabla S(x, t)$, or the so-called "guidance condition", as can be easily seen by writing $\psi = R \exp[\mathrm{i}S]$. A similar argument using

$$T^{00} = \frac{\mathrm{i}}{2} [\psi^* \partial^0 \psi - \psi \partial^0 \psi^*]$$

gives the Bohm energy, $E_B(x, t) = -\partial_t S(x, t)$. Thus the Bohm energy and momentum are nothing but the energy–momentum density derived from the standard expression for the energy–momentum tensor. Hiley and Callaghan (2012) have shown that similar relationships hold for the Pauli and Dirac particles.

In the standard approach the energy, E, is defined as

$$E(t) = \int T^{00}(x, t) \mathrm{d}^3 x \qquad (6.11)$$

while the momentum P^j is defined by

$$P^j(t) = \int T^{0j}(x, t) \mathrm{d}^3 x. \qquad (6.12)$$

These quantities are not a function of position but are global quantities. Thus the conventional theory is concerned with the *global* energy and momentum, whereas the Bohm approach focuses on the *local* expressions, $E_B(x, t), P_B^j(x, t)$. Conservation of global energy and momentum is achieved through

$$\frac{\mathrm{d}}{\mathrm{d}t} \int T^{0\mu}(x, t) \mathrm{d}^3 x = 0, \qquad \mu = 0, 1, 2, 3.$$

As we have already pointed out in Section 6.1, conservation of local energy is achieved using the quantum Hamilton–Jacobi Equation (6.2) (see Hiley and Callaghan, 2012). The differences between these two approaches to the notion of particle number has been discussed by Colosi and Rovelli (2009).

6.3.3 Weak values and the $T^{0\mu}(x, t)$ components of the energy–momentum tensor

We will now show how the $T^{0\mu}(x, t)$ components of the energy–momentum tensor are related to weak values. To do this we take Equation (6.10) and replace the operator $\hat{A}$ with the momentum operator $\widehat{P}_x$, while replacing $\langle\phi|$ by $\langle x|$. This means we are investigating the weak value of the momentum at a post-selected position x. Then

$$\langle\psi(t)|\widehat{P}_x|\psi(t)\rangle = \int \rho(x,t)\frac{\langle x|\widehat{P}_x|\psi(t)\rangle}{\langle x|\psi(t)\rangle}\mathrm{d}x.$$

Now we can write

$$\langle x|\widehat{P}_x|\psi(t)\rangle = \int \langle x|\widehat{P}_x|x'\rangle\langle x'|\psi(t)\rangle\mathrm{d}x'.$$

However, we use $\langle x|\widehat{P}_x|x'\rangle = -\mathrm{i}\nabla_x\delta(x-x')$ again, so that

$$\langle x|\widehat{P}_x|\psi(t)\rangle = -\mathrm{i}\nabla_x\psi(x,t).$$

If we now write $\psi(x,t) = R(x,t)\mathrm{e}^{\mathrm{i}S(x,t)}$ we find

$$\frac{\langle x|\widehat{P}_x|\psi(t)\rangle}{\langle x|\psi(t)\rangle} = \nabla_x S(x,t) - \mathrm{i}\nabla_x\rho(x,t)/2\rho(x,t),$$

where $\rho(x,t) = |\psi(x,t)|^2$ is the usual probability density. If we now form a bilinear invariant of the second kind by writing

$$\frac{\langle x|\overleftrightarrow{P}_x|\psi(t)\rangle}{\langle x|\psi(t)\rangle} = \nabla_x S(x,t), \tag{6.13}$$

we see that this invariant is, in fact, the $T^{01}(x,t)$ component of the energy–momentum tensor which also happens to be the x-component of the Bohm momentum (Leavens, 2005; Wiseman, 2007). The imaginary part of the weak value is the osmotic velocity which is generally ignored, but see Bohm and Hiley (1989) for further details.

If we formally introduce the time operator $\widehat{P}_t$, the real part of the weak value $\langle P_t\rangle_W$ gives $T^{00}(x,t)$ and hence the Bohm energy, while the real part of $\langle P_t^2\rangle_W$ gives the Bohm kinetic energy plus the quantum potential (Leavens, 2005). The corresponding expressions for a particle with spin has been presented in Hiley (2012).

Having seen the relevance of weak values, we now need to show how they can be realized experimentally.

6.4 Weak measurements with photons

6.4.1 The experiment of Kocsis et al.

Weak measurements have already been made on very weak photon beams in a beautiful experiment by Kocsis *et al.* (2011a) (see Figure 6.1). Polarized single photons are split by a 50-50 beam splitter before being re-coupled using two collimated fiber couplers that act as two slits. After the photon has passed through the screen containing the two slits, the real part of the weak value of the transverse momentum is measured at various points. From these values a set of stream lines are constructed which have a striking similarity to the trajectories calculated by Philippidis, Dewdney and Hiley (1979) for a Schrödinger particle described by the Bohm model (see Figure 6.2).

The weak measurement of the momentum was made by introducing a thin slither of birefringent calcite crystal into the photon beam immediately after the two slits. The calcite was thin enough and placed with its optical axis suitably oriented so that the ordinary and extraordinary ray were still overlapping after they left the crystal. This process induces a phase change in the beam that can be calculated using the interaction Hamiltonian $\widehat{H}_I = g\widehat{P}_x\widehat{S}_1$, where $2\widehat{S}_1 = [|H\rangle\langle H| - |V\rangle\langle V|]$. Here $|H\rangle$ and $|V\rangle$ are the horizontal and vertical components of the polarization of the photon. This gives the final state of the photons to first order to be

$$\Psi(x_f) \approx \frac{\psi(x_f)}{\sqrt{2}}\left[e^{-i\phi}|H\rangle + e^{i\phi}|V\rangle\right],$$

where we have written $\phi = \frac{D\Delta t}{2}\langle P_x\rangle_W$, the theoretical value of the phase. In the Kocsis *et al.* experiment, the phase was written as $\phi = \zeta\langle P_x\rangle_W$ where ζ is some factor that depends on the details of the experimental setup, as discussed in Kocsis *et al.* (2011b).

To measure this phase factor, we need to perform a strong measurement using $S_3 = |R\rangle\langle R| - |L\rangle\langle L|$. Here $|R\rangle$ and $|L\rangle$ are the right and left circularly polarized states respectively. Then we find that

$$\langle S_3\rangle = \sin[\zeta\langle P_x\rangle_W].$$

The weak value of the transverse momentum can then be found from the difference in the counts of the number of right-hand photons and left-hand photons arriving at a point through the relation

$$\langle P_x\rangle_W = \frac{1}{\zeta}\sin^{-1}\left(\frac{I_R - I_L}{I_R + I_L}\right).$$

This then determines the transverse momentum at a series of points in the interference region. From these values, together with the momentum along the axis, Kocsis *et al.* (2011a) construct the momentum stream lines, as explained in detail in their paper.

6.4.2 The meaning of the stream lines

The question that must now be addressed is precisely what meaning can be attached to these stream lines in the case of photons. Since we are here dealing with the electromagnetic field, a weak measurement of the momentum will give values of the Poynting vector of this field at various points after the field has passed through the thin calcite crystal used to perform the weak measurement. The experiment of Kocsis *et al.* (2011a) is dealing with single photons, posing the question "What is the meaning of the Poynting vector when a single photon is involved?"

To make the discussion more transparent, let us consider first a measurement that would determine the $T^{00}(x.t)$ component, namely, the electromagnetic energy. If the beam was a monochromatic plane wave composed of many photons, we would argue, along with Dirac (1927), that if we divide this value by the number of photons, we could determine the energy carried by a single photon.

This type of argument has already been used in the Beth (1936) experiment to measure mechanically the spin of a photon. In this experiment a circularly polarized beam of light is reflected back by a mirror after passing through a quartz wave plate. If the quartz wave plate is fixed to the mirror and the combined system attached to a quartz fiber, the change of the beam's angular momentum in the wave plate at reflection will produce a torque in this fiber. The change of angular momentum taking place in the quarter-wave plate induces a torque on the quartz wave plate, causing it to rotate through an angle that enables the change in angular momentum of the reflected beam to be measured. When one divides this value by the number of photons in the beam, we find that each photon carries a unit of angular momentum as expected.

Since in the Kocsis *et al.* experiment only a single photon enters the apparatus at a time, the weak measurement of the energy will give a value for the energy carried by a single photon. A corresponding measurement of the momentum will give us the momentum of a single photon.

We are now faced with an interesting question: "What is the meaning of the energy and momentum of a photon at a point?" The photon has a sharply defined energy and momentum so it could be argued that the uncertainty principle would forbid us the talk about a photon at a sharply defined position. Detailed discussions of these problems can be found in Cook (1982), Mandel (1983) and Roychoudhuri, Kracklauer and Creath (2008).

In the Bohm approach, sharp "values" can be simultaneously attributed to an object (the beables), but if we try to attribute a simultaneous position and momentum to a photon while maintaining the field values for the energy, $T^{00} = (E^2 + B^2)/8\pi$, and momentum, $T^{0i} = (B \times E)^i/4\pi$, we run into trouble. This problem has already been pointed out by Bohm, Hiley and Kaloyerou (1987) and by Bohm and Hiley (1993), but its significance does not seem to have been appreciated so we will briefly outline the problem using a simple example.

Consider the special case in which B is parallel to E so that $E \times B = 0$. Assume further that E and B are in the x^1 direction. This means that under Lorentz boost in the x^1 direction, we will still have E' parallel to B' so that $(E' \times B') = 0$. It follows that the velocity of the photon is zero in a whole range of Lorentz frames, which makes little sense. Indeed, no meaning can be given to a photon at rest in general. This gives rise to serious doubts as to whether one can think of the photon as a particle even in the Bohmian sense.

Indeed, this was one of the reasons why Bohm (1952b), Bohm, Hiley and Kaloyerou (1987) and Bohm and Hiley (1993) proposed that photons, and bosons in general, should be treated in terms of a field. This meant developing a Bohmian approach to quantum field theories (Bohm *et al.,* 1987). This theory treats the field $\phi(x^\mu)$ and its conjugate momentum $\pi(x^\mu)$ as the beables. The dynamics is introduced in terms of a super-Schrödinger equation, which, in turn, leads to a super-quantum Hamilton–Jacobi equation together with a super-quantum potential which organizes the fields to produce the quantum effects. We will not discuss this approach here but the details for a scalar field can be found in Bohm, Hiley and Kaloyerou (1987) while those for the electromagnetic field can be found in Kaloyerou (1994). A simplified account can also be found in Bohm and Hiley (1993).

For our purposes here, all we need to know is how the concept of a photon arises in Bohm's approach to the field. When an atom emits a quantum of energy, the energy disperses by spreading into the global electromagnetic field so that the energy of the field increases by $h\nu$. This energy is not localized and this is reflected in standard quantum field theory where an integral over all space is performed as shown in Equations (6.11) and (6.12).

The atom absorbs energy from the field through the action of the super-quantum potential. This potential is non-local and sweeps up a quantum of energy, $h\nu$, from the field, causing an appropriate local transition in the atom. It is the structure of the energy levels in the atom that determines the amount of energy absorbed or emitted. More details can be found in Bohm, Hiley and Kaloyerou (1987) and Kaloyerou (1994).

This is a radically different way of understanding the concept of a photon from the usual one, although even in the more conventional case there is no clear view as to how we are to understand the physical nature of the photon (see Roychoudhuri, Kracklauer and Creath, 2008). Not only does it avoid the difficulties we have discussed above, but it also offers an explanation for the coherent state where, although the quantum state of the field is well-defined, the number of photons it contains is not. If photons existed as well-defined actual entities then it is hard to understand why the number of entities in a definite state should be ambiguous. A further advantage to this approach is that the non-local nature of the quantum potential gives an account of the EPR effect, as was shown in detail in Bohm *et al.* (1987).

Returning to the photons considered in the experiment of Kocsis *et al.*, we note that the individual photons are described by a *Gaussian wave packet*, indicating that the photons are not mono-energetic, a small spread of frequencies is involved. In order to find the weak value of the momentum at any point, we must place a detector at that point and it is by counting photons that the weak values are determined. So in terms of this model, the stream lines that Kocsis *et al.* (2011a) construct are simply energy–momentum flow lines and do not imply that photons actually follow these flow lines.

6.4.3 Schrödinger particle trajectories

The objections that we have raised against photon trajectories do not apply to atoms. If we assume that atoms can be described by the Schrödinger equation, then, as we have already pointed out, the Bohm approach allows us to calculate trajectories in a two-slit interference experiment, as was shown by Philippidis *et al.* (1979).

An atom, with its finite rest mass, can be brought to rest and can thus have a simultaneous position and momentum, even though we will not be able to measure these values simultaneously using a von Neumann-type strong measurement. However, as we have seen in Section 6.3.3, this simultaneous momentum is the Bohm momentum, which is not an eigenvalue of the momentum operator in the state under consideration, but the weak value defined in Equation (6.13). This means that this momentum can be measured by making the appropriate weak measurement of the momentum operator. Since atoms move relatively slowly, the question of Lorentz invariance will not arise, so that we can discuss the notion of a particle trajectory without encountering the problems that beset the notion of a photon trajectory. Thus in this case a measurement of the appropriate weak values will allow us to construct an ensemble of trajectories which can then be compared with the theoretical calculations of Philippidis *et al.* (1979).

6.5 Conclusions

In this chapter we have discussed how weak measurements give new information about quantum processes. In particular we have shown how weak values give new information about spin and about energy and momentum. In fact we have shown that a weak value of the energy and momentum operators using position as a post-selection gives us values for the $\{0, \mu\}$ components of the energy–momentum tensor $T^{\mu\nu}(x, t)$. In turn we showed that the Bohm energy, $E_B(x, t)$, and the Bohm momentum, $P_B(x, t)$, are simply related to the appropriate values of $T^{0\mu}(x, t)$. Thus values for these Bohm variables can be obtained by the measurement of the corresponding weak values, showing that the Bohm approach is not a return to classical notions. It gives us information about the local energy and momentum, whereas the standard interpretation gives us information about their global properties.

To date the weak measurement techniques described in this chapter have been applied only to optical systems such as Young's two slit experiment (Kocsis *et al.,* 2011a) and the optical analogue of the classic Stern–Gerlach experiment (Ritchie *et al.,* 1991). What has not yet been demonstrated is the use of weak measurements for particles with non-zero rest mass obeying the Schrödinger equation. Results here will enable us to make a direct comparison with the theoretical predictions of the Bohm model.

With this objective in mind we propose to carry out a series of experiments starting with a modified SG apparatus shown in Figure 6.4. We plan to confirm that the amplification factors that are measured correspond to those predicted. We then plan to apply the method to a modified Young's slits apparatus aiming to reproduce the "photon trajectories" found by Kocsis *et al.* (2011a), only in our case we will use atoms. This will open the possibility of employing the technique to observe weak effects assumed to be beyond our reach using conventional techniques. In this way we believe that we can make experiments that observe quantum effects that were once thought to be impossible.

Acknowledgments

The authors would like to thank J. Clarke and K. Marinov of the Daresbury Laboratory and Cockroft Institute for providing much needed expertise in magnet design. We would like to thank the HEP group at UCL for its support thus far and any support it can provide into the future. We would also like to thank Taher Gozel for his generous financial support and encouragement.

References

Aharonov, Y., Albert, D. Z. and Vaidman, L. (1988) How the result of a measurement of a component of the spin of a spin-1/2 particle can turn out to be 100, *Phys. Rev. Lett.,* **60**, 1351–1354.

Aharonov, Y. and Vaidman, L. (1990) Properties of a quantum system during the time interval between measurements, *Phys. Rev. A,* **41**, 11–19.

Aharonov, Y. and Vaidman, L. (1993) Measurement of the Schrödinger wave of a single particle, *Phys. Lett. A,* **178**, 38–42.

Aharonov, Y. and Vaidman, L. (1995), Protective measurements, *Annals of the New York Academy of Sciences,* **755**, 361373. doi: 10.1111/j.1749-6632.1995.tb38979.x

Beth, R. A. (1936) Mechanical detection and measurement of the angular momentum of light, *Phys. Rev.,* **50**, 115–125.

Bohm, D. (1951) *Quantum Theory*, Englewood Cliffs, Prentice-Hall, NJ.

Bohm, D. (1952a) A suggested interpretation of the quantum theory in terms of hidden variables, I, *Phys. Rev.*, **85**,166–179.

Bohm, D. (1952b) A suggested interpretation of the quantum theory in terms of hidden variables, II, *Phys. Rev.*, **85**, 180–193.

Bohm, D. and Hiley, B. J. (1989) Non-locality and locality in the stochastic interpretation of quantum mechanics, *Phys. Reps.*, **172**, 92–122.

Bohm, D. and Hiley, B. J. (1993) *The Undivided Universe: an Ontological Interpretation of Quantum Theory*, London, Routledge.

Bohm, D., Hiley, B. J. and Kaloyerou, P. N. (1987) An ontological basis for the quantum theory: II – a causal interpretation of quantum fields, *Phys. Rep.*, **144**, 349–375.

Colosi, D. and Rovelli, C. (2009) What is a particle? *Classical and Quantum Gravity*, **26**, 025002.

Cook, R. J. (1982) Photon dynamics, *Phys. Rev. A*, **25**, 2164–67.

Dirac, P. A. M. (1927) The quantum theory of the emission and absorption of radiation, *Proc. Roy. Soc.*, **114A**, 243–265.

Duck, I. M., Stevenson, P. M. and Sudarshan, E. C. G. (1989) The sense in which a "weak measurement" of a spin-1/2 particle's spin component yields a value 100, *Phys. Rev. D*, **40**, 2112–17.

Dürr, D., Goldstein, S. and Zanghi, N. (1996) Bohmian mechanics as the foundation of quantum mechanics, in *Bohmian Mechanics and Quantum Theory: an Appraisal*, J. T. Cushing, A. Fine and S. Goldstein (eds.), Boston Studies in the Philosophy of Science, **184**, 21–44, Dordrecht, Kluwer.

Heisenberg, W. (1949) *The Physical Principles of Quantum Mechanics,* trans. by C. Eckart and F. C. Hoyt, New York, Dover.

Heisenberg, W. (1958) *Physics and Philosophy: the Revolution in Modern Science*, London, George Allen and Unwin.

Hiley, B. J. (2012) Weak values: approach through the Clifford and Moyal algebras, *J. Phys.: Conference Series*, **361**, 012014. doi: 10.1088/1742-6596/361/1/012014.

Hiley, B. J. and Callaghan, R. E. (2010a) The Clifford algebra approach to quantum mechanics A: the Schrödinger and Pauli particles. arXiv:Maths-ph:1011.4031.

Hiley, B. J. and Callaghan, R. E. (2010b) The Clifford algebra approach to quantum mechanics B: the Dirac particle and its relation to the Bohm approach. arXiv: Maths-ph:1011.4033.

Hiley, B. J. and Callaghan, R. E. (2012) Clifford algebras and the Dirac and Bohm quantum Hamilton–Jacobi equation, *Foundations of Physics*, **42**, 192–208. doi: 10.1007/s10701-011-9558-z.

Hosoya, A. and Shikano, Y. (2010) Strange weak values, *J. Phys. A: Math. Theor.* **43**, 385307.

Kaloyerou, P. N. (1994) The causal interpretation of the electromagnetic field, *Phys. Rep.*, **244**, 287–358.

Kocsis, S., Braverman, B., Ravets, S., *et al.* (2011a) Observing the average trajectories of single photons in a two-slit interferometer, *Science*, **332**, 1170–1173.

Kocsis, S., Braverman, B., Ravets, S., *et al.* (2011b) Supporting online material, www.sciencemag.org/cgi/content/full/332/6034/1170/DC1.

Leavens, C. R. (2005) Weak measurements from the point of view of Bohmian mechanics, *Found. Phys.*, **35**, 469–491. doi: 10.1007/s10701-004-1984-8.

Mandel, L. (1983) Photon interference and correlation effects produced by independent quantum sources, *Phys. Rev. A*, **28**, 929–943.

Philippidis, C., Dewdney, D. and Hiley, B. J. (1979) Quantum interference and the quantum potential, *Nuovo Cimento*, **52B**, 15–28.

Ritchie, N. W., Story, J. G. and Hulet, R. G. (1991) Realization of a measurement of a weak value, *Phys. Rev. Lett.*, **66**, 1107–1110.

Roychoudhuri, C., Kracklauer, A. F. and Creath, K. (2008) (eds.) *The Nature of Light: What is a Photon?* Boca Raton, CRC Press.

Schweber, S. S. (1961) *An Introduction to Relativistic Quantum Field Theory*, New York, Harper-Row.

Takabayasi, T. (1955) On the structure of Dirac wave function, *Prog Theor. Phys.*, **13**, 106–108.

von Neumann, J. (1955) *Mathematical Foundations of Quantum Mechanics*, Princeton, Princeton University Press.

Wiseman, H. M. (2007) Grounding Bohmian mechanics in weak values and Bayesianism, *New J. Phys.*, **9**, 165–177. doi: 10.1088/1367-2630/9/6/165.

Part II
Meanings and implications

7

Measurement and metaphysics

PETER J. LEWIS

7.1 Introduction

It is a prima facie reasonable assumption that if a physical quantity is measurable, then it corresponds to a genuine physical property of the measured system. You can measure a person's mass because human beings have such a property. You can measure the average mass of a group of people because groups of people have such a collective property. And so on.

Now it would be truly surprising – miraculous, perhaps – if you could determine the average mass of a group of people by making measurements on just one of them. To ascribe such a statistical property to an individual looks like a category mistake. At first glance, protective measurements seem to pull off just such a miracle, determining, for example, the expectation value of position for an ensemble of particles via a measurement performed on one of them. The lesson we are supposed to draw, of course, is that expectation values are not statistical properties at all, despite their name. Rather than being an average over an ensemble of systems, the expectation value of position for a particle is a physical property of the individual system, and the wave function, as the bearer of these properties, is a physical entity (Aharonov, Anandan and Vaidman, 1993).

The protective measurement procedure has been challenged (Uffink, 1999; Gao, 2013; Uffink, 2013), but for present purposes I will assume that protective measurements exist, at least in principle, that are capable of revealing "statistical" properties like expectation values in a single measurement. My aim here is not to challenge the existence of such a physical procedure, but rather to explore the arguments that connect the existence of protective measurements with conclusions concerning the nature of physical reality. What protective measurements are supposed to show is that "epistemological" interpretations of the quantum state are untenable – that the wave function of a system must instead be interpreted "ontologically" (Aharonov, Anandan and Vaidman, 1993: 4617).

But what exactly are the epistemological and ontological interpretations contrasted here? There are at least two distinct possibilities.[1] First, the epistemological interpretation could be identified with an empiricist attitude towards quantum mechanics in general – taking the theory as a recipe for generating the probabilities of measurement results. Here the contrast is with scientific realism, construed as the view that quantum mechanics is in some sense a true description of the physical world. However, Dickson (1995) has convincingly argued that protective measurement cannot decide between empiricism and realism about quantum mechanics, since protective measurement is entirely consistent with empiricism.[2] Hence I set this construal of the argument aside.

Second, the contrast between the epistemological and ontological interpretations of the wave function could be construed within an overall scientific realist attitude towards quantum mechanics in terms of the distinction between a statistical description and a categorical description of the physical system in question. Under this construal, the ontological interpretation is that the wave function is a description of the properties of a single physical entity, whereas the epistemological interpretation is that the wave function is a description of the distribution of properties over an ensemble of similar physical systems.

Taken in this way, the argument in favor of the ontological interpretation is more interesting. At first glance, though, it is still somewhat puzzling, since the conclusion of the argument is already the standard position among realist interpretations of quantum mechanics, prior to any consideration of protective measurement. No-go theorems (e.g. Bell, 1964; Kochen and Specker, 1967) are taken to show that it is impossible to interpret the wave function as a statistical distribution of properties over an ensemble of similar systems. Interference phenomena seem to show that elements of the wave function physically interact with each other, and hence that the wave function is something like a physically real field. Indeed, all three of the major realist research programs in the foundations of quantum mechanics – Bohmian, Everettian and GRW – take the wave function to be a real dynamical entity.[3]

Perhaps the intention is just to put the last nail in the coffin of ensemble interpretations. However, my goal here is to show that even though ensemble interpretations face formidable (and well-known) obstacles, protective measurements don't lead to any additional difficulties. Rather, they provide us with a nice illustration

---

[1] Aharonov, Anandan and Vaidman take the target of their argument to be the position that "the wave function represents at least partially our knowledge of the system" (1993: 4617), but this isn't decisive between the two construals I outline here.

[2] Dickson does contend, though, that protective measurement puts realism and empiricism "back on an even footing", since it counteracts the support for instrumentalism provided by ordinary impulsive measurement (1995: 135).

[3] However, it is worth noting that some Bohmians argue that the wave function is a physical law rather than a physical entity (Dürr, Goldstein and Zanghì, 1996).

of a conclusion for which we had considerable indirect evidence already, namely that quantum mechanics leads to a blurring of the distinction between the intrinsic properties of a system and the statistical properties of the ensemble of which it is a member. This conclusion goes for all realist interpretations of the quantum state, both the mainstream ones that take the wave function to be a real field and the more conjectural ones that take the wave function to describe our knowledge of an ensemble.

7.2 Bohm's theory

Although the usual target of arguments from protective measurement is ensemble interpretations of the wave function, protective measurements have also been used to argue against the tenability of Bohm's theory (Aharonov, Englert and Scully, 1999). The argument against Bohm's theory begins from a protective measurement that measures the wave function intensity in a small region around $x = 0$ for a particle whose wave function is spread out along the x axis. Wave function intensity, of course, is connected via the Born rule to the statistical properties of the system; in this case, it tells us the probability of finding the particle close to $x = 0$ if we measure its position using a standard impulsive measurement. The fact that the wave function intensity itself can be directly measured seems to show that the wave function intensity is a physical property of the system.

By itself, this is no argument against Bohm's theory, since Bohm's theory already takes the wave function intensity to be a real physical property of the system; it is the wave function that pushes the particle around. Rather, the argument concerns the physical details of the measurement interaction. The protective measurement is effected by a Hamiltonian that links the state of the measured particle with the state of a second particle that acts as a pointer. In particular, the Hamiltonian contains a weak, long-lasting coupling term in the region around $x = 0$ between the two particles. This interaction induces a shift in the wave packet of the pointer particle proportional to the wave function intensity of the measured particle in the $x = 0$ region.

What (supposedly) makes this interaction problematic for Bohm's theory is that the Bohmian particle configuration for the system never passes through the $x = 0$ region. The interaction Hamiltonian is such that the measured particle and the pointer particle only interact within the $x = 0$ region. So the Bohmian particle configuration for the system cannot represent the actual particle motions during a protective measurement. Protective measurements, so the argument goes, cannot be given a physically acceptable Bohmian analysis.

I think this argument involves a misunderstanding of the role of particles in the Bohmian theory. Bohmian particles are not dynamically active entities; they do not act on each other or on the wave function. Their role is passive and phenomenological; they are pushed around by the wave function and correspond to the observed outcomes of our measurements. So the fact that the particle configuration does not pass through the $x = 0$ region is irrelevant to the explanation of the dynamical behavior of the system, since the particles are dynamically inert. All the action lies in the evolution of the wave function, construed as a real field. The wave function surely exists (in part) in the $x = 0$ region, and the dynamical behavior of the wave function in this region, under the influence of the Hamiltonian, explains the motion of the wave packet for the pointer particle. The only role of the Bohmian particle configuration is to pick out one precise location in this wave packet as representing the observed outcome of the measurement.[4]

The confusion arises, I suspect, because the Hamiltonian itself is derived from an analogous interaction in classical mechanics, and in the classical case the interaction really does occur between particles at $x = 0$. In classical mechanics, the particles are dynamically active – the locus of forces. Classically, if the particles did not pass through the interaction region, the interaction could not occur. The Bohmian particles take on the phenomenological role of the classical particles; their configuration corresponds to the observed outcome of a measurement. But they do not take on the dynamical role of the classical particles; that role is taken on by the wave function.

It is instructive to compare the protective measurement to standard two-slit interference, in which the particle passes through one slit or the other and the wave function passes through both slits. Suppose the Bohmian particle passes through the left-hand slit. Its trajectory is affected by whether or not the right-hand slit is open, even though the particle doesn't pass through the right-hand slit. One might try to parlay this into an objection to Bohm's theory – how could the particle be affected by the state of a slit it doesn't go near? – but clearly this would be unfair to the theory. The particle trajectory is not intended to be a dynamical explanation, as it is in classical mechanics.

This is not to say that Bohm's theory is unproblematic. In systems with more than one particle, the motion of one particle depends on the location of all the other particles, no matter how distant. This makes it hard to square Bohm's theory with special relativity, since the location of distant particles *right now* is undefined in special relativity. This problem certainly arises in protective measurements, since they involve more than one particle, but it is hardly a special problem for protective measurements, and it was well-known before protective measurements were postulated (e.g. Bell, 1971).

---

[4] Indeed, this minimal role for the particle configuration has been exploited by Everettians seeking to argue that the Bohmian particles are redundant (e.g. Brown and Wallace, 2005).

7.3 Contextual properties

So Bohm's theory suffers from no special problem explaining protective measurements, but only generic problems of an explicitly non-local theory. And Bohm's theory is particularly useful in making explicit the distinction between contextual and non-contextual properties of quantum systems. Since this distinction will be important later, it is worth considering in detail how contextual properties arise in Bohm's theory.

A contextual property is one whose value depends on the context in which it is measured. The x-spin of a particle prepared in an eigenstate of z-spin provides a simple example of a contextual property in Bohm's theory.[5] Suppose the x-spin of the particle is measured using a Stern–Gerlach device; if the particle is spin-up it is deflected upwards, and if it is spin-down it is deflected downwards. Since Bohm's theory is a deterministic theory, the possible trajectories of the Bohmian particles cannot cross in configuration space (as an intersection between trajectories would amount to indeterminism). In a single-particle experiment like that envisioned here, the configuration space is just ordinary three-space. This means that if the particle starts out in the upper half of the wave packet, it is deflected upwards, and if it starts out in the lower half it is deflected downwards (because otherwise the trajectories would cross). So (one might think), spin-up particles are just those that are located in the upper half of the wave packet.

But now suppose we rotate the Stern–Gerlach device by 180° around the paths of the particles, so that spin-down particles are deflected upwards and spin-up particles are deflected downwards. Consider again a particle that starts out in the upper half of the wave packet; by the same argument as before, it must be deflected upwards. But now this particle is recorded as a spin-down particle. That is, a particle that starts out in the upper half of the wave packet will manifest itself as a spin-up particle if the Stern–Gerlach device is one way up, but a spin-down particle if the device it the other way up. The spin of the particle depends on the context in which it is measured.

This way of putting things makes it sound as if spin in Bohm's theory is a property of particles, albeit of a peculiar kind. But this is somewhat misleading. In fact, rather than saying that spin is a (contextual) property of the measured system, it is more accurate to say that the measured system doesn't really have a spin property – that nothing in its pre-measurement state corresponds to the observed spin value. After all, the orientation of the Stern–Gerlach device can't affect the earlier state of the measured system, so prior to the measurement there can be no fact of the matter about whether the system has the spin-up property or the spin-down property. However, contextuality doesn't arise for a particle residing in a wave packet that is

---

[5] My exposition here follows Albert (1992).

an eigenstate of x-spin. In such a case, the entire wave packet is deflected up (say), taking the particle with it, and if the Stern–Gerlach device is rotated, the entire wave packet is deflected down, again taking the particle with it. Here it does seem appropriate to ascribe a spin property to the system. And presumably the bearer of this spin property is the wave function, since, as noted above, the particles are causally inert in Bohm's theory. So all in all, it seems that spin is a property that the *wave function* possesses in Bohm's theory, but only when the wave function takes certain forms (the spin eigenstates).

Similar considerations apply to the protective measurement described in the previous section. In this measurement, the wave packet of the measured particle is confined within a box centered on $x = 0$ in the ground state (Aharonov, Englert and Scully, 1999: 138). The protective measurement effectively measures an observable for which this state is an eigenstate whose eigenvalue is the wave intensity around $x = 0$ (Uffink 1999: 3479). The measurement outcome is recorded in the position of the pointer particle, which let us suppose moves along its y-axis.

But now suppose that the wave packet of the measured particle is initially a superposition of the ground state and the first excited state. This wave packet is not an eigenstate of the observable measured by the protective measurement; the ground state has large wave intensity around $x = 0$, but the first excited state has small wave intensity here. (For this reason the measurement would not count as protective, since it disturbs the measured system.) Hence the pointer particle ends up in a superposition of two distinct positions; its wave function has significant intensity in two distinct regions, close to $y = 0$ and far from $y = 0$. If the pointer wave function is initially concentrated close to $y = 0$, then the wave function splits into two packets, one of which moves up and the other of which stays put. Consider the motion of Bohmian particles in this case. Since Bohmian trajectories cannot cross, if the particle starts in the upper half of the initial wave packet it moves up, and if it starts in the lower half it stays put.[6] Now consider what happens if the direction of the measurement Hamiltonian is reversed; the term in the pointer superposition corresponding to the ground state moves down rather than up, so particles that start in the lower half of the initial wave packet move and those in the upper half stay put. For a given initial configuration of the Bohmian particles, then, whether the measurement indicates that the wave intensity close to $x = 0$ is small or large depends on the orientation of the measuring device.

So, just like x-spin, the property measured by this protective measurement is contextual according to Bohm's theory. And again, since the state of the measured

---

[6] Matters are complicated slightly here by the fact that we are dealing with a two-particle system inhabiting a six-dimensional configuration space. Bohmian trajectories cannot cross in this configuration space. However, since the Bohmian trajectories are practically stationary in the coordinates of the measured particle (Aharonov, Englert and Scully, 1999: 144), this amounts to the condition that the trajectories do not cross in the coordinates of the pointer particle.

system can't depend on measurements that have yet to be performed on it, to say that a property is contextual is tantamount to saying that it doesn't exist at all; the pre-measurement system lacks a property corresponding to wave intensity at the origin. As before, this contextuality doesn't arise when the state of the measured particle is an eigenstate of the measured observable (so that the measurement is genuinely protective). In that case, it is appropriate to ascribe a wave intensity property to the system, where the bearer of the property is the wave function rather than the particle. That is, the wave intensity measured by this protective measurement is a property that the wave function sometimes possesses – a property that it possesses when it is in the ground state, but does not possess when it is a superposition of the ground and first excited states.

One might object to this conclusion on the grounds that the measurement just described is not a protective measurement (as noted above), since it disturbs the measured system. A genuine protective measurement is available that measures the wave intensity around the origin for the superposition state; this requires a different Hamiltonian to ensure that the measurement is non-disturbing.[7] On the basis of this genuine protective measurement, one might reasonably infer that since the wave intensity can be measured in the superposition state, the wave function possesses a wave intensity property for this state too. However, it is equally true that one can measure the spin property of a particle in a superposition of distinct x-spin eigenstates without disturbing its state. Suppose the particle is initially in a symmetric superposition of x-spin eigenstates. Then one can measure the spin property of this particle without disturbing it by rotating the Stern–Gerlach device by 90°, since the symmetric superposition of x-spin eigenstates is an eigenstate of spin along the z-axis. But of course, although the particle has a z-spin property in this state, it does not have an x-spin property (and we know this because an x-spin measurement on the z-spin eigenstate is contextual).

A similar thing can be said in the case of wave intensity measurements. Although one can perform a non-disturbing measurement that returns a wave intensity value for the superposition state, the corresponding property that this state possesses cannot be the same as the property measured by the original protective measurement on the ground state. Whatever property the original protective measurement measures, the superposition of ground and protective states lacks it, since the measurement on this state is contextual. The superposition state has a related property, measured by the new protective measurement, one that perhaps has an equal claim to be called "wave intensity". Hence wave intensity, like spin, is best thought of as

---

[7] It might be argued that the only difference in the Hamiltonian between the two measurements is the protective term that ensures that the measured state remains undisturbed. That is, the term in the Hamiltonian that defines the measured property remains unchanged. But it would need to be shown that a principled distinction can be drawn between these various parts of the Hamiltonian – one that does not beg the question by assuming that the measured property remains the same.

a family of related properties, not a single property, and different protective measurements measure different wave intensity properties.

The upshot of this argument is that although Bohmians typically take the wave function to be a physical entity, its property structure is quite complex. In particular, the intensity of the wave function around a particular point in (configuration) space is perfectly well defined in the theory, but can generate disparate measurement results depending on how it is measured, indicating that it should not be viewed as a single possessed property of the system. That is, although the wave function underlies the physical properties of a system, in the sense that a difference in physical properties requires a difference in the wave function, it does not follow that all the mathematical properties of the wave function correspond to simple physical properties of the system. The contextual nature of wave intensity in Bohm's theory explored above suggests that even though wave intensity can be measured via a protective measurement, it should not be regarded as a simple physical property of the system.

7.4 Ensemble interpretations

In the previous section I argued that we have to be careful in our inference that the result of a protective measurement corresponds to a property of the measured system. So far, though, nothing I have said challenges the basic argument that protective measurements show that the wave function is a physical entity, and hence rule out ensemble interpretations of the wave function. Even though the wave intensity property can be contextual, there is no contextuality when the state is an eigenstate of the protective measurement operator. So the existence of protective measurements shows that there exist measurements that reveal "statistical" properties like wave intensity, and reveal them with certainty without disturbing the measured system. Hence we can turn Einstein's criterion of reality against him. Einstein famously held that "if, without in any way disturbing a system, we can predict with certainty (i.e., with probability equal to unity) the value of a physical quantity, then there exists an element of reality corresponding to that quantity" (Einstein, Podolsky and Rosen, 1935: 777). Since protective measurements allow us to predict the values of wave function properties with certainty (and without disturbing the system), there must be elements of reality corresponding to these properties, undermining Einstein's own hope that the wave function can be interpreted as a statistical description of an ensemble rather than a physical entity.

As noted above, however, ensemble interpretations were apparently ruled out by the no-go theorems long before the advent of protective measurements. Still, there are those who continue to hold out hope. One prominent strategy for evading the no-go theorems is to invoke retrocausality – the hypothesis that causal influences can travel from later events to earlier events, as well as in the usual fashion from earlier events to later events (Cramer, 1986; Kastner, 2013; Price, 1994; Price and Wharton, 2013; Sutherland, 2008; Wharton, 2010). How does this strategy fare against the challenge of protective measurement?

It is important to note that contextuality is typically ubiquitous in retrocausal theories. Since causal influences can travel from later measurement events to earlier system preparation events, it is always possible for the initial state of a system to reflect the measurements that will later be performed on it.[8] Unlike the contextuality that emerges in Bohm's theory, retrocausal contextuality can in principle even affect systems in eigenstates of the measured observable. For example, suppose that the state of a one-particle system is prepared in the spin-up eigenstate of x-spin. Then if the x-spin of the system will later be measured, the possessed properties of the system must correspond to x-spin-up, in the sense that those possessed properties will bring about the spin-up result with certainty. But if the x-spin of the particle will not later be measured, this fact can affect the earlier possessed properties of the particle, so they might differ from those that correspond to x-spin-up.

Consider what this means for the protective measurement of wave intensity discussed above. Like Bohm's theory, retrocausal theories ascribe determinate trajectories to a set of particles, but unlike Bohm's theory, there is no wave function steering the particles. So retrocausal theories cannot avail themselves of the response to Aharonov, Englert and Scully I gave on behalf of Bohm's theory; since there is no wave function in a retrocausal theory, the particles themselves must be the dynamically active entities. A further disanalogy with Bohm's theory is that we cannot rule out the possibility of particle trajectories crossing in the retrocausal case; since we have as yet no explicit dynamics for a retrocausal theory, we cannot know whether a fully formulated retrocausal theory would be deterministic. But the crucial disanalogy with Bohm's theory is that any property can be contextual in a retrocausal theory, even the position of a particle, and even if the quantum state is an eigenstate of the observable to be measured.

Given this last point, it immediately follows that the property measured by a protective measurement can be contextual according to a retrocausal theory. That is, the possible particle trajectories can be sensitive to the kind of measurement that will be performed on the system. If a protective measurement is performed, this later fact can (in principle) cause the earlier particle trajectories to all pass through the region around $x = 0$, and hence deflect the pointer particle via the interaction Hamiltonian. If, on the other hand, a standard impulsive measurement is performed on the system, this later fact can (in principle) cause the earlier particle trajectories

---

[8] Indeed, this is precisely how retrocausal theories evade the no-go theorems (Price, 1994).

to be statistically distributed according to the wave function intensity, so that the wave function can be interpreted epistemically. In the latter case, presumably only a small proportion of the particle trajectories pass through the region around $x = 0$. (The caveat "in principle" is a huge one here, of course, since nobody has succeeded in explicitly constructing such a theory.)

What should we make of the ontological status of contextual properties in retrocausal theories? In Bohm's theory, I suggested that contextual properties are strictly speaking not properties of the system at all, since the state of the system now cannot depend on the measurements that will be performed on it later. But in a retrocausal theory, the state of the system now *can* depend on the measurements that will be performed on it later, so there is no barrier to contextual properties being genuine possessed properties of the system. And since there is no wave function in a retrocausal theory, these contextual properties have to be possessed properties of the *particles* in the system.

There is something a little strange in this, perhaps: the "wave intensity" property measured by the protective measurement is in fact the property of a particle, not a wave. But it is surely no stranger than "particle" properties like spin turning out to be properties of the wave function in Bohm's theory, GRW and Everett. In GRW and Everett "particles" are really just manifestations of the wave function; the wave function mimics particles under certain circumstances. Even in Bohm's theory, particle properties other than position are carried by the wave function. Retrocausal theories manifest the opposite effect; particles sometimes mimic the properties we might otherwise attribute to the wave function. The ontology underlying a particular measurement result depends on the theory used to explain that result; it can't be read off the measurement itself.

One might object that the fact that the particles in retrocausal theories have properties like "wave intensity" that can be measured using protective measurements means that the wave function in effect still exists, since wave-like properties are still physically instantiated. But it is worth noting that the contextuality of retrocausal theories means that only those "wave-like" properties whose values will be determined by protective measurements on the system are actually instantiated. A typical system subject only to impulsive measurements will have no such properties. A system subject to the above protective measurement will have a property corresponding to wave intensity at the origin, but not (for example) to the expectation value of position. In non-retrocausal theories like Bohm's theory the state cannot depend on which measurements will be performed on the system, and so all possible wave function properties must be instantiated at once. Only in this latter case must we postulate the wave function as a physical entity.

Of course, there are formidable obstacles to be faced by any ensemble interpretation of the quantum state. While retrocausal theories have long been proposed, nobody has yet succeeded at explicitly formulating the dynamics for such a theory. Perhaps it will prove to be impossible. My point here is only that protective measurements provide no *new* argument against ensemble interpretations; any interpretation that can bypasses the no-go theorems by appealing to contextual properties can thereby also evade the protective measurement argument for the reality of the wave function.

7.5 Ensemble properties and individual properties: a blurring of the lines

Suppose an ensemble interpretation of quantum mechanics along the lines given in the previous section is possible. Then protective measurements can be explained without postulating a genuinely wave-like entity; the motion of the particle can fully explain the result. But there is still an air of mystery surrounding this account of protective measurement. The result of the protective measurement lines up with a genuinely statistical property – the proportion of particles one *would* find close to $x = 0$ were one to perform a series of ordinary impulsive measurements on an ensemble of similarly prepared particles. What explains this agreement, if the protective measurement itself just records the value of a possessed property of the particle? How can the properties of a single particle reflect the statistical properties of an ensemble of particles?

What I wish to suggest is that this agreement is somewhat mysterious for *all* the major realist traditions for interpreting quantum mechanics, so there is nothing special about ensemble interpretations in this regard. The Bohmian, Everettian and GRW programs also explain the result of the protective measurement in terms of a possessed property of the system, albeit a property of the wave function. So these theories also face the problem of how a possessed property of an individual system can reflect the statistical properties of an ensemble of similarly prepared systems. Postulating an entity that is spread out in configuration space as the bearer of the property does not solve the problem of how the actual wave-intensity property reflects the statistical properties of sets of unactualized impulsive measurements.

The mystery is deepest in Everettian quantum mechanics. There are no particles in the Everettian theory, so the result of the protective measurement described above is explained by a property of the wave function. What if an ordinary impulsive measurement is performed on the same system? Then the Schrödinger dynamics drives the wave function into a set of (practically) non-interacting branches, one for each possible outcome. The wave intensity of these branches matches the wave intensity of the corresponding elements of the original measured state – so, for example, if an impulsive measurement is performed to determine whether the particle is close to $x = 0$, the intensity of the "yes" branch will be the same as the wave intensity of the measured state close to $x = 0$. But now comes the tricky part: it needs to be established that the intensity of a branch corresponds to its probability. There is a well-known research program seeking to establish this (Deutsch, 1999; Wallace, 2012), but it remains controversial. There is certainly no *obvious* explanatory link between the intensity of a branch – which is a possessed property of a single entity – and the frequency of a given outcome in an ensemble. If the Deutsch–Wallace program succeeds, it does so by uncovering a *surprising* explanatory link here; if it does not succeed, it is because there is no such link at all.

Bohmian and GRW theories are built on the assumption that the Everettian explanation fails, and both attempt to add such an explanation by modifying quantum mechanics. The GRW collapse law is a stochastic law tailored specifically to connect wave intensity with the probability of recording the associated result were an ordinary impulsive measurement to be performed on the state. That is, the connection between the possessed properties of the wave function and the statistical properties of impulsive measurements is established by fiat. Whether one finds this genuinely explanatory probably depends on one's general feelings about the explanatory status of brute propensities (Dorato and Esfeld, 2010). Bohmian interpretations use much the same strategy. The Bohmian dynamical law is formulated specifically to ensure that the particle positions are always statistically distributed according to wave function intensity, hence connecting wave function intensity with the statistics of ordinary impulsive measurements in the required way. Again, this connection is established by fiat.

In this context, the explanation embodied by a putative ensemble interpretation does not look so bad after all. The explanation for the agreement between the result of the protective measurement and the statistical distribution of impulsive measurement results is that the retrocausal dynamics takes all the particle trajectories through the $x = 0$ region in the protective case, but distributes them so that only a small proportion of them pass through this region in the impulsive case. Whether the dynamical law that accomplishes this shares the *ad hoc* flavor of the GRW and Bohmian dynamical laws remains to be seen; as yet we have no such law. It is promising, I think, that retrocausal theories are being developed based on the Feynman path construction, in which a quantum system probes all possible paths between two points (Wharton, Miller and Price, 2011). This makes it more plausible that a substantive explanation might be found whereby the present state of the system reflects the statistical properties of future possibilities. But even if no such explanation is forthcoming and the law establishes the link by fiat, it is in good company.

The main virtue of the literature on protective measurement, it seems to me, is to bring to the fore the rather remarkable connection between ensemble properties and individual properties in quantum mechanics. Prior to quantum mechanics, one would have said that applying a statistical property like an expectation value to an individual system is just a category mistake; statistical properties only properly apply to ensembles of identically prepared systems. But the postulation of the wave function as a physical entity by Everettian, Bohmian and GRW quantum mechanics means that statistical properties like expectation values are reflected in the actual properties of a single system. Protective measurements do not provide a new argument for this conclusion, but they make it manifest in a remarkably direct way.

But the link between the existence of protective measurements and the existence of the wave function as a physical entity should not be overstated. *Some* property of the measured system must correspond to the result of the protective measurement, but the protective measurement itself provides no evidence that the property must be instantiated by a wave-like entity rather than a particle. That is, measurement alone can't tell you what exists; rather, you have to look at the best *theoretical explanation* of the measurement results, and infer what exists from the ontological commitments of that theory. It is true that wave function explanations are dominant at the moment, but they are far from problem-free, and retrocausal particle-only explanations remain a promising alternative.

Acknowledgements

I would like to thank Shan Gao and the participants in the Workshop in Epistemology and Philosophy of Science at the University of Miami for very helpful comments on an earlier draft of this chapter.

References

Aharonov, Y., Anandan, J. and Vaidman, L. (1993). Meaning of the wave function. *Physical Review A*, **47**, 4616–4626.

Aharonov, Y., Englert, B. G. and Scully, M. O. (1999). Protective measurements and Bohm trajectories. *Physics Letters A*, **263**, 137–146.

Albert, D. Z. (1992). *Quantum Mechanics and Experience*. Cambridge, MA: Harvard University Press.

Bell, J. S. (1964). On the Einstein–Podolsky–Rosen paradox. *Physics*, **1**, 195–200. Reprinted in Bell (1987).

Bell, J. S. (1971). Introduction to the hidden-variable question. *Foundations of Quantum Mechanics: Proceedings of the 49th International School of Physics "Enrico Fermi"*. New York: Academic, pp. 171–181. Reprinted in Bell (1987).

Bell, J. S. (1987). *Speakable and Unspeakable in Quantum Mechanics*: Cambridge: Cambridge University Press.

Brown, H. R. and Wallace, D. (2005). Solving the measurement problem: de Broglie–Bohm loses out to Everett. *Foundations of Physics*, **35**, 517–540.

Cramer, J. G. (1986). The transactional interpretation of quantum mechanics. *Reviews of Modern Physics*, **58**, 647–687.

Deutsch, D. (1999). Quantum theory of probability and decisions. *Proceedings of the Royal Society of London*, **A455**, 3129–3137.

Dickson, M. (1995). An empirical reply to empiricism: protective measurement opens the door for quantum realism. *Philosophy of Science*, **62**, 122–140.

Dorato, M. and Esfeld, M. (2010). GRW as an ontology of dispositions. *Studies in History and Philosophy of Modern Physics*, **41**, 41–49.

Dürr, D., Goldstein S. and Zanghì, N. (1996). Bohmian mechanics and the meaning of the wave function. In *Experimental Metaphysics: Quantum Mechanical Studies in Honour of Abner Shimony*, R. S. Cohen, M. Horne and J. Stachel (eds.). Dordrecht: Kluwer, pp. 25–38.

Einstein, A., Podolsky, B. and Rosen, N. (1935). Can quantum-mechanical description of physical reality be considered complete? *Physical Review*, **47**, 777–780.

Gao, S. (2013). On Uffink's criticism of protective measurements. *Studies in History and Philosophy of Modern Physics*, **44**, 513–518.

Kastner, R. E. (2013). *The Transactional Interpretation of Quantum Mechanics*. Cambridge: Cambridge University Press.

Kochen, S. and Specker, E. P. (1967). The problem of hidden variables in quantum mechanics. *Journal of Mathematics and Mechanics*, **17**, 59–87.

Price, H. (1994). A neglected route to realism about quantum mechanics. *Mind*, **103**, 303–336.

Price, H. and Wharton, K. (2013). Dispelling the quantum spooks – a clue that Einstein missed? arXiv:1307.7744.

Sutherland, R. I. (2008). Causally symmetric Bohm model. *Studies in History and Philosophy of Modern Physics*, **39**, 782–805.

Uffink, J. (1999). How to protect the interpretation of the wave function against protective measurements. *Physical Review A*, **60**, 3474–3481.

Uffink, J. (2013), Reply to Gao's "On Uffink's criticism of protective measurements". *Studies in History and Philosophy of Modern Physics*, **44**, 519–523.

Wallace, D. (2012). *The Emergent Multiverse: Quantum Theory According to the Everett Interpretation*. Oxford: Oxford University Press.

Wharton, K. (2010). A novel interpretation of the Klein–Gordon equation. *Foundations of Physics*, **40**, 313–332.

Wharton, K. B., Miller, D. J. and Price, H. (2011). Action duality: a constructive principle for quantum foundations. *Symmetry*, **3**, 524–540.

8

Protective measurement and the explanatory gambit

MICHAEL DICKSON

Quantum theory has traditionally – and not altogether unreasonably – been taken as a challenge to "realism" about physical theories. At the very least, the ways in which it is often formulated, presented and used suggest a non-realist understanding of the theory because the significance of the "state" of a system is described in terms of a catalogue of predictions. While protective measurement does not force one to give up on this standard view, it does support an alternative contention, namely, that the state of a physical system (a wave function, for example) has a somewhat more direct physical significance. I conclude with what I take to be the upshot of these observations for approaches to interpreting quantum theory and evaluating those interpretations.

8.1 Introduction

In 1993 I wrote an article (Dickson, 1995) about the scheme for protective measurement first described (to my knowledge) by Aharonov, Anandan and Vaidman (1993). There I claimed that protective measurement makes "realism" about quantum theory more attractive than it might otherwise have been. I still believe some form of that claim to be true, and I am grateful to the editor and to Cambridge University Press for the opportunity to say so in the manner that I'm inclined to now, 20 years later.

Of course, the debates about the interpretation of quantum theory have moved on quite a bit in the intervening time. We've seen a rise to prominence of "sub-jectivist" interpretations, largely based on developments in quantum information theory (but with its own historical precedents, as Timpson (2010) has pointed out – see Fuchs (2002) for a landmark paper in the modern development, and Timpson (2013) for a thorough overview and evaluation). The various "many-somethings" (worlds, minds, perspectives, whatever) interpretations have, arguably, seen a rise in fortune as well. Nonetheless, while there may be a shift of emphasis in the contemporary discussions, the fundamental challenge to interpretation has remained steady: as it is frequently used and formulated, quantum theory seems to beg for some form of non-realist (subjectivist, instrumentalist, empiricist) interpretation, and yet often those interpretations have a hard time making sense of the fact that quantum theory is *also* naturally taken to be about – and is ultimately measured against experimental determinations of – the way the physical world is, independently of us, our instruments, or our epistemological scruples.

In my view, protective measurement gives a gentle nudge away from facile subjectivist, instrumentalist, or empiricist interpretations of quantum theory, though it does no damage to more careful and sophisticated versions of those views. It does, however, pose a challenge to them, insofar as it reminds us in a very precise way that quantum theory is, or at any rate certainly appears to be, somehow, about "how the world is" (including parts of the world not within the ken of the empiricist), an appearance that must be explained by these interpretations.

In the next section, I say a bit more carefully what I take some of these various positions (realism, empiricism, etc.) to be, and also say something more about how quantum theory has traditionally been understood to bear on them (an understanding that I more or less share). Then, in the subsequent section, I review the idea of protective measurement, and point out where it suggests a slight change to this traditional understanding. In the final section, I step back and evaluate the situation, saying something more general about the challenges facing any attempt to "interpret" quantum theory.

8.2 Realisms and non-realisms

There are too many approaches to quantum theory (and nuances to those approaches) to review them all here. I will choose a few from a spectrum of possibilities, as representatives of the whole.

The first position is "realism", which, as I shall use the term, is the position that *some* appropriate version of quantum theory should be understood as making literally true, or approximately true, statements about the physical world. There are, in other words, physical things (or classes of them) that serve as referents for the "objects" that appear in the theory, and those physical things (or members of the relevant classes) have properties that correspond more or less directly to the predicates that appear in (some appropriate version of) the theory. Realists take one of two approaches. Either they provide a formulation of the theory that differs from the standard formulations used by physicists, and argue for (or in any case, adopt) a realist position as regards that alternative formulation, or they accept a standard formulation and accept the apparently bizarre consequences of a realist stance towards it.

For example, Bohmians (see, e.g., Bohm and Hiley, 1993, or Dürr, Goldstein, and Zanghì, 1997) and advocates of the various "dynamical collapse" theories (see, e.g., Ghirardi, Rimini and Weber, 1986, or Pearle, 1989) provide non-standard formulations, in an explicit attempt to find a theory whose objects and predicates might plausibly be understood in a realistic manner, typically, though not necessarily, describing some form of mass or energy distributed in space and evolving over time in a manner consistent with the predictions of standard quantum theory. Whether they succeed on physical terms and whether they succeed at being plausibly realistic are of course matters of debate, but whether their approach is more or less as described here is, I think, undisputed.

My description is not meant to be pejorative in any way. To some, "altering" quantum theory is some sort of sin against the theory that is to be avoided if possible. (The claim is often made by advocates of the many-somethings approach, alongside the claim that the many-somethings approach adds nothing to the theory.) I see no reason to take this view, but if you do, and if you have some form of realist tendency, you might be inclined towards the many-somethings interpretations, which purport to leave quantum theory "untouched" and realistically interpreted. (It isn't clear to me that they do either, but for an extended argument to the contrary, see, e.g., Wallace (2012). Indeed, I'm a bit skeptical that the very idea of an interpretation's "leaving the theory untouched" makes much sense.) On this view, the state is taken to represent "reality", though in the end it is not entirely clear (to me) what that "reality" consists. Advocates of the many-somethings views suggest that decohered "branches" of the state of a system (ultimately, the Universe) correspond to "worlds" that are (or at least, some of which are) more or less the way a naïve realist might suppose, but as far as I can tell the view leaves unspecified any account of how we are to understand the (ontological) nature of the collection of such worlds, much less the nature of a system that has not experienced decoherence (and none of them has, perfectly), apart from the use of colorful language such as "multiverse".

So, in the end, I'm not quite sure where to place the many-somethings interpretations as regards protective measurement, the most natural account of which involves (so I shall argue) the assertion that there is some sort of causal interaction between something that answers to "the wave function of the system" and an apparatus. On the other hand, in Bohm's theory and in the dynamical collapse theories, it is at least superficially clear what it would mean for such a causal interaction to occur, i.e., how it is possible.

In any case, all of these realist positions have costs, and from a non-realist perspective, the costs are ultimately attributable to the misguided attempt to understand quantum theory as describing a "way the world is". Here I consider three such approaches.

On the subjectivist approach, quantum theory is (primarily) about "information", or "subjective degrees of belief" (these two things being sometimes identified, sometimes not). In one form, as described (and later criticized) by Timpson, the view is that

The quantum state ascribed to an individual system is understood to represent a compact summary of an agent's degrees of belief about what the results of measurement interventions on a system will be, **and nothing more**.

(Timpson, 2008, section 2.1, emphasis original.)

The view is not *entirely* non-realist about the theory as a whole – some room is left for the theory to be telling us *something* about "how the world is", but the quantum state, at least, is understood in this entirely subjectivist way. ("Subjectivist", here, is being used in the sense of the subjectivist interpretation of probability.)

The constructive empiricist approach, due primarily to van Fraassen (1980, 1991), takes the view that the aim of science is not literal truth (or approximate truth) about all of the matters under its purview. Rather, science aims at truth only about directly observable matters of empirical fact, a goal van Fraassen calls "empirical adequacy". Now, there is plenty to ponder here, about the limits of observation, but it is clear that for van Fraassen, the quantum state of, say, a hydrogen atom, is not among those things that are directly observable. Instead, van Fraassen offers a "modal" interpretation of the state.

It is important, for van Fraassen, to offer *some* understanding of the quantum state, because according to constructive empiricism, we are to take the theoretical claims about unobservables literally. (They are not, for example, elliptical claims about sense data, as some positivists might have held, or about measuring apparatuses and procedures, as some early operationalists might have held.) Hence there is real work for the constructive empiricist to do, namely, to spell out this literal understanding of quantum states. Van Fraassen's motivating idea

is to deny neither the determinism of the total system evolution [apparatus plus measured system] nor the indeterminism of outcomes, but to say that the two are different aspects of the total situation. Specifically, we can deny the identification of value-attributions to observables with attributions of states; the state can then develop deterministically, with only statistical constraints on changes in the values of the observables.

(van Fraassen, 1991, 273.)

While there are subtleties to consider, the main result of this approach is that the significance of the quantum state is twofold. First, it is used to determine the (completely deterministic) dynamics of a system (in the usual way, via the Schrödinger equation). Second, it is used (via an algorithm that need not concern us) to determine which properties a system *could* have at any moment. At least in typical measurements, the prescription for this determination is designed to yield a set of possible properties that seems to correspond with the properties that one is inclined to suppose are being probed by the measurement. Quantum theory is understood as ascribing, probabilistically, one of these possible properties to the system.

The instrumentalist approach to quantum theory is in agreement with van Fraassen about the aim of science (empirical adequacy rather than literal truth) but does not agree about the literal understanding of state-ascriptions (and other theoretical claims). Instead, the instrumentalist views the theory – or some parts of the theory – as tools, in a straightforward sense. They are built (or found), like tools, and used for a specific (in this case, predictive or explanatory) purpose, and are not to be read as *descriptions* of the world. One who is instrumentalist about the quantum state, then, does not take it to be a description of the world, but a tool used for predictive or explanatory (or perhaps some other) theoretical purpose. In its most common form, instrumentalism takes the quantum state to be a kind of dynamical catalogue of the possible results of possible experiments (and their probabilities). Hence, whatever else it might be, it is not a description of a property of a thing in the world with which one might interact. (Philosophers of physics tend to dismiss instrumentalism as silly, or already refuted long ago; they also tend to suppose that the instrumentalist automatically has a solution to the interpretive problems of quantum theory. Both moves are a mistake, in my view, but I will not, here, attempt to develop a more interesting (and vulnerable) version of instrumentalism.)

Each of these non-realist approaches (and to some extent Everettian many-somethings approaches as well) is traditionally motivated by and framed in terms of the standard, simplified, account of measurement, where we consider a system in a state, $|\psi\rangle$, and the measurement of a physical property represented by a self-adjoint operator, $\mathbf{F}$, and it is supposed that what quantum theory tells us about "what it is to be in the state $|\psi\rangle$" is that, after the measurement, the system will be found (by us) to be in one of the eigenstates, $|f_i\rangle$ of $\mathbf{F}$, each with probability $|\langle\psi|f_i\rangle|^2$. Or at any rate, each of these interpretations has traditionally been aimed at trying to make sense of the undeniable and widespread success of that understanding of "what it is to be in the state $|\psi\rangle$".

But what if there is *another* understanding of what it is to be in the state $|\psi\rangle$, one that arises quite naturally from the quantum formalism itself and does *not* appear to be describable in terms of ascribing one of several possible properties to the system (probabilistically)? Then it is an open question how the various approaches – realist, subjectivist, constructivist, instrumentalist – will fare in their attempts to make sense of it.

8.3 Protective measurement

Indeed there *is* an alternative (or perhaps better, supplementary) understanding of "what it is to be in the state $|\psi\rangle$", a possibility brought to light by Aharonov, Anandan and Vaidman (1993), in the form of so-called "protective measurement". The basic idea is simple: it may be possible to "protect" the wave function of a system from collapsing during a measurement. If so, then, by appropriate measurements, we can begin to build up a picture of the wave function itself via a series of interactions that, for all the world, look like they are best understood as interactions between the measurement apparatus and the wave function of the measured system.

It is easy to forget (and I think that until the appearance of Aharonov, Anandan and Vaidman's paper, many of us *had* forgotten) that the standard account of measurement is based on (or at any rate, is most plausibly justified, theoretically, in terms of) a particular model of a measurement interaction, the so-called "impulsive" model. In this model, the exchange of energy between the apparatus and the system is described in terms of an interaction Hamiltonian, and the broad-brushed features of the interaction are characterized by the duration and strength of the interaction. The standard idea (as worked out, for example, in London and Bauer (1939), Bohm (1951), and von Neumann (1955)) is that the interaction is very short and very strong, the so-called "impulsive" model of measurement. In contrast, Aharonov, Anandan, and Vaidman (1993) consider measurement interactions that are (relatively) long, and weak.

Let's begin with a quick review of the standard idea. We write the total Hamiltonian for a (to be) measured system and the measuring apparatus as

$$H_{\text{total}} = H_{\text{S}} + H_{\text{M}} + H_{\text{I}},$$

where H_{S} is the (free) Hamiltonian for the (to be measured) system, H_{M} is the (free) Hamiltonian for the measuring apparatus, and H_{I} is the interaction Hamiltonian (between the two systems). Let $H_{\text{I}} = g(t)(P \otimes A)$, where P is an observable of the apparatus (the momentum observable, as it will turn out), A is an observable of the (to be measured) system, and $g(t)$ is a function that characterizes the strength of the interaction as a function of time. One simple way to proceed is to let $g(t) = \gamma f(t)$, where γ is a constant that characterizes the "overall strength" of interaction, and $f(t)$ characterizes the degree of coupling over time. As an approximation, we say that $f(t)$ is zero everywhere except inside some region of interaction, $[0, \tau]$, and without loss of generality we presume that $f(t)$ is normalized, so that $\int_0^\tau f(t)\mathrm{d}t = 1$.

Let the initial total state of the compound system be $|\Psi(0)\rangle = |\psi(0)\rangle |\chi(0)\rangle$ (leaving the tensor product implicit and letting ψ refer to the (to be) measured system, and χ to the apparatus). If we presume that γ is "large" and τ is "small", then the Schrödinger equation together with standard mathematical manipulations yield a final state for the compound system, at the time τ, of

$$|\Psi(\tau)\rangle = \sum_k \langle a_k|\psi(0)\rangle \, |a_k\rangle \, |\xi_k\rangle ,$$

where the $|a_k\rangle$ are the eigenstates of A (presumed, here, for simplicity, to be non-degenerate) and the $|\xi_k\rangle$ are states of the apparatus given by

$$|\xi_k\rangle = e^{-i\gamma a_k P} \, |\chi(0)\rangle .$$

That is, they are spatial translations of the apparatus (a "pointer"), each by an amount γa_k. (See Dickson, 1995, for a slightly more careful and more general account, and for references to texts that deal with the technical matters.)

This "impulsive" model yields the standard account of measurement, in which the combined system ends up in a superposition of compound states in each of which (under a natural interpretation) the apparatus indicates a value, a_k, and the measured system is in the corresponding eigenstate of A, $|a_k\rangle$. And thus, so the standard story goes, $|\psi\rangle$ provides a "catalogue" of possible results of the measurement, and assigns a probability ($|\langle\psi|a_k\rangle|^2$) to each. Once the apparatus is found to be in the state $|\xi_k\rangle$, then, according to standard usage, the entire system "collapses" to the state $|a_k\rangle |\xi_k\rangle$. This "collapse", upon measurement, of the state to just one of the terms merely reflects which item in the catalogue turned up.

Let's contrast that scheme for measurement interactions with the scheme for protective measurements. The set-up is the same, but now we will consider what happens to the compound system when γ is small and τ is large. Let's suppose, as well, that $|\psi(0)\rangle$ is the ground state of a one-dimensional harmonic oscillator, whose Hamiltonian is therefore H_S. (I will return to the role and status of this assumption later.) Let's now take the interaction Hamiltonian to be

$$H_I = \gamma f(t) Q \otimes A, \tag{8.1}$$

where Q is the position of the apparatus (e.g., a "pointer" position) and A is again some observable of the system (and henceforth I leave tensor products implicit). The solution to the Schrödinger equation for times, $\tilde{t}$, just after τ is

$$\left|\Psi(\tilde{t})\right\rangle = \exp\left[-i(H_S\tilde{t} + H_M\tilde{t} + \gamma QA)\right] \langle a_k|\psi(0)\rangle |a_k\rangle |\chi(0)\rangle , \tag{8.2}$$

where the a_k are the eigenvectors of A. Appealing to an argument made by Schiff (1968, 289–291), if γ is small enough, the energy imparted to the measured system (essentially by the term $\exp[-i\gamma QA\tilde{t}]$, though some care is required because the largeness of τ means that we cannot ignore the other terms) during the interaction is not additive, and is thus adiabatically negligible – it will not excite the measured system out of its ground state. So we need consider the effect of the evolution operator on $|\chi(0)\rangle$, which yields (after a little manipulation)

$$|\Psi(t)\rangle \approx \sum_k |\psi(0)\rangle |\xi_k(t)\rangle, \qquad (8.3)$$

where

$$|\xi_k(t)\rangle = \exp\left[-\mathrm{i}(H_M\tilde{t} + \gamma QA)\right]|\chi(0)\rangle. \qquad (8.4)$$

An application of the Schrödinger equation yields

$$\frac{\mathrm{d}}{\mathrm{d}t}\langle\chi(t)|P|\chi(t)\rangle = -g(t)\langle\psi(0)|A|\psi(0)\rangle, \qquad (8.5)$$

so that we can use the "pointer momentum" of the apparatus (which could be, for example, a probe particle) to measure $\langle\psi(0)|A|\psi(0)\rangle$. Equation (8.3) shows that after this measurement, the state of the measure system is unchanged; therefore, I'll just write it as $|\psi\rangle$.

So let A be the projection, P_Δ, onto a small region of space. Then our protective measurement reveals $\langle\psi|P_\Delta|\psi\rangle$, which, in a position representation (i.e., write $|\psi\rangle$ as a wave function) is

$$\int_\Delta \psi^*(x)\psi(x)\mathrm{d}x, \qquad (8.6)$$

i.e., the number that we would normally associate with the *probability* that the system would be found in the region Δ, were we to look for it there. By measuring P_Δ for various regions, Δ, we could, in principle, begin to build up a spatial picture of the (modulus-squared of the) wave function. We could even then go on to do the very same thing with observables that do not commute with position. (There is no violation of the Uncertainty Principle – quantum theory predicts that if our picture of the wave function in a position representation is sharply peaked, for example, then it will be found to be spread out in a momentum representation.)

There are a few observations to make here. First, Aharonov, Anandan and Vaidman (1993) offered some other means of "protecting" the state of the system during measurement. I will not comment on those other methods here.

Second, the scheme is not foolproof. The adiabatic approximation is just that, an approximation. There will always be a non-zero probability that something goes wrong (e.g., that the system is excited out of its ground state, in our example). (See, e.g., Dass and Qureshi, 1999.) Nothing in what I say here or later is damaged by this observation, for the only point that matters, for now, is that a protective measurement *could* (according to quantum theory) be made, successfully.

Third, in the scheme discussed here, one *does* need to *know* the state of the measured system prior to measuring it. That knowledge would presumably come from a preparation or a measurement (of, say, the energy of the harmonic oscillator), and presumably the correct model for such a preparation or measurement is the impulsive model. But then what has the subsequent protective measurement really told us about the state of the measurement system that we did not already know? Nothing! So is it really even a measurement?

I will not quibble (much) over terminology, for it matters not, for my point, whether the subsequent manipulations of the system "count" as a genuine measurement. (Were I to quibble, I would observe that one could, for example, put a system in the ground state then tell somebody else that the state (unknown this second observer) is "protected". That person could then discover the state, and surely what he or she does is "discovery by measurement".) What matters for me, here, is that it appears, for all the world, that over the course of the series of protective measurements, the "thing in the world" with which we are interacting, and whose properties are being discovered, is the wave function. The "catalogue" view of the "meaning" of the wave function does not appear to be particularly illuminating in this case. So let us turn, finally, to that point and address it in slightly more detail.

8.4 The explanatory gambit

The "collapse upon measurement" that is part of the standard account of (impulsive) measurement should, and does, get realists worked up, for a number of familiar reasons, the most fundamental being that, for a realist, "what happens in the theory" is supposed to reflect, somehow, "what happens in the world"; and it is hard to see how traditional collapse of this sort *could* happen, physically. (It violates the basic dynamics of the theory; it is inherently discontinuous, which is normally a recipe for physical disaster; it wreaks havoc on any attempt to describe the world in a manner that is consistent with relativity theory; and it's just downright odd, depending, as it does, on the potentially problematic notion of "measurement".) As I suggested earlier, realistic interpretations of the theory are motivated by the desire to escape this problem.

And escape it they do, in various ways that I rehearsed earlier. But there is always a cost. These costs are familiar from discussions of the merits (or lack of merits) of various interpretations, which I will not rehearse here. (Much of what I said on these points in Dickson (1998) remains true; of course, plenty of additional discussion has occurred since then.) The fact is that "realist" interpretations, in various ways, stretch credulity in ways that call into question at least their explanatory power, if not their coherence.

Consider this example. In Bohm and Hiley's (1993) version of Bohm's theory, the quantum potential (which is responsible for the specifically quantum behavior of particles) has a veneer of classicality. For example, it figures in the equation of motion in more or less the same way that classical potentials do. But underneath, the quantum potential has some pretty bizarre properties. For example, it has no source. It is, in a fairly robust sense, non-local. And more. Bohm and Hiley are of course willing to embrace these properties and take them as discoveries. Others are more apt to apply *modus tollens*. To them, explanation of non-local correlation in terms of the quantum potential amounts to reducing the perplexing to the mysterious.

In my view (but I will not make the case here), all realistic interpretations face the same issue: they are committed to an account of the physical world that many will find unexplanatory, in the sense that the proposed explanans is at least as much in need of explanation as the explanandum.

Considerations such as these can easily drive one to some form of a non-realist interpretation. While quite different in their commitments, the non-realist interpretations are in agreement about a *lack* of commitment in realist-style explanations of quantum behavior. In particular, if one is not wedded to the idea that the theory straightforwardly describes how the world is – and therefore one is not (necessarily) bothered if something that happens "in the theory" does not look like something that could plausibly happen "in the world" – then one might not be bothered by "collapse", at least in the sense that one might not be bothered by *explaining* it. (One might still be worried about formulating a more precise version of the collapse postulate, or even concerned to eliminate it from the theory. Van Fraassen, for example, seems to have a concern in this vicinity.)

However, this is not to say that non-realists are not committed to explanation in some form. For example, the general form of an instrumentalist explanation of the success of a physical theory is that the theory was *designed* to be a successful tool (and insofar as humans are not too terrible at making tools, it is to be expected that a concerted effort to make a tool that performs a given task – such as predicting the outcomes of experiments – will succeed[1]). Indeed, it is hard to see what the point of a non-realist interpretation would be if not to *understand* something about physical theories, including their success. Moreover, given the close connection between understanding and explanation, one would therefore expect that non-realists have a concern with explaining *something* about quantum theory.

Given that there is *some* concern with explanation (and if there is no such concern, it is hard to see how the non-realist is playing anything like the "interpretation game"), the problem that is posed for non-realists by protective measurement is as follows. As I mentioned above, non-realist accounts are motivated by the "catalogue" view of the quantum state, and they employ some form of it in their account of quantum theory, and therefore in their account of how and why quantum

---

[1] The preceding sentence might sound like an endorsement of the view that "explaining X" amounts to "making X seem expected or unsurprising". I think that there is something to this notion, but do not mean to be endorsing it as an *account* of explanation.

theory generates the predictions it does for various measurements. Moreover, the catalogue account makes good sense of the generic non-realist contention that we should not understand measurement in terms of some realistic and causal account of the interaction between a measuring apparatus and the properties of the measured system. To put it the other way around, the non-realist has a pretty good explanation ("pretty good" from his or her own perspective) of the standard quantum theoretic account of measurement in terms of the "catalogue" afforded by the quantum state: quantum theory invokes the catalogue account of the quantum state because to invoke something else (such as a causal interaction between the apparatus and properties of the measured system) would belie a misunderstanding of the (non-realist!) nature of science.

But protective measurement throws a wrench into these works. The catalogue account of the quantum state does a terrible job of explaining what is going on in a protective measurement. Consider again the sequential (protective) measurement of P_Δ (for different Δ), discussed in the previous section. Each one appears, for all the world, to be the measurement of a feature of the "actual" wave function of a single system, and whatever items in a catalogue might be, they do not appear to be capable of explaining such an appearance.

Of course, the non-realist can always retreat and point out that at *some* point, an impulsive measurement will be made (for example, of the momentum of the pointer), and the *rest* of the "internal" interactions are just "some stuff that happens" in the theory, not to be characterized as a measurement. Such a position is of course logically available, as is a retreat from the catalogue view in favor of a less specific form of non-realism. However, in both cases, the non-realist's account becomes less explanatorily powerful. In contrast, one who is realist about the quantum state is not only not bothered by protective measurement, but takes glee in the fact that it appears very amenable to being explained in terms of a causal interaction between real properties of a quantum system (its quantum state) and a measuring apparatus.

So what's the upshot? There is no easy answer. This, I contend, is the explanatory gambit in quantum theory, and it arises in a variety of specific forms: interpretations that make good sense of one aspect of the theory tend to do poorly on others. For a time (or so I contend), non-realist interpretations had something of an upper hand. (As a matter of sociological fact, amongst philosophers of physics, it seems that a majority were realist, but this fact (if true) says more about a widespread commitment to realism than it does about the goodness of fit between realism and quantum theory.) Protective measurement – especially if its implementation in the laboratory becomes widespread – goes a good way towards redressing this balance, not (to my mind) to the point of giving realist interpretations the upper hand, but to the point of leveling the playing field.

References

Aharonov, Y., Anandan, J. and Vaidman, L. (1993), Meaning of the wave function, *Physical Review*, **47**: 4616–4626.

Bohm, D. (1951), *Quantum Theory*. New York: Prentice-Hall.

Bohm, D. and Hiley, B. (1993), *The Undivided Universe: an Ontological Interpretation of Quantum Theory*. London: Routledge and Kegan Paul.

Dass, N. D. H. and Qureshi, T. (1999), Critique of protective measurements, *Physical Review A*, **59**: 2950.

Dickson, M. (1995), An empirical reply to empiricism: protective measurement opens the door for quantum realism, *Philosophy of Science*, **62**: 122–140.

Dickson, M. (1998), *Quantum Chance and Nonlocality*. Cambridge: Cambridge University Press.

Dürr, D., Goldstein, S. and Zanghì, N. (1997), Bohmian mechanics and the meaning of the wave function, in Cohen, R. S., Horne, M. and Stachel, J. (eds.), *Experimental Metaphysics – Quantum Mechanical Studies for Abner Shimony*, Volume One. Boston Studies in the Philosophy of Science, **193**. Boston: Kluwer Academic Publishers.

Fuchs, C. A. (2002), Quantum mechanics as quantum information (and only a little more), in Khrenikov, A. (ed.), *Quantum Theory: Reconsideration of Foundations*. Växjö, Sweden: Växjö University Press.

Ghirardi, G. C., Rimini, A. and Weber, T. (1986), Unified dynamics for microscopic and macroscopic systems, *Physical Review*, **D34**: 470.

London, F. and Bauer, E. (1939), *La théorie de l'observation en mécanique quantique*. Actualites Scientifiques et Industrielles, vol. 775. Paris: Hermann et Cie.

Pearle, P. (1989), Combining stochastic dynamical state-vector reduction with spontaneous localization, *Physical Review*, **A39**: 2277.

Schiff, L. (1968), *Quantum Mechanics*. 3rd edn. New York: McGraw-Hill.

Timpson, C. G. (2008), Quantum Bayesianism: a study, *Studies in History and Philosophy of Modern Physics*, **39**: 579–609.

Timpson, C. G. (2010), Information, immaterialism, instrumentalism: old and new in quantum information, in A. Bokulich and G. Jaeger (eds.), *Philosophy of Quantum Information and Entanglement*, 208–228. Cambridge: Cambridge University Press.

Timpson, C. G. (2013), *Quantum Information Theory and the Foundations of Quantum Mechanics*. Oxford: Oxford University Press.

van Fraassen, B. (1980), *The Scientific Image*. Oxford: Oxford University Press.

van Fraassen, B. (1991), *Quantum Mechanics: an Empiricist View*. Oxford: Oxford University Press.

von Neumann, J. (1955), *Mathematical Foundations of Quantum Mechanics*. Trans. R. T. Beyer. Princeton: Princeton University Press.

Wallace, D. (2012), *The Emergent Multiverse: Quantum Theory According to the Everett Interpretation*. Oxford: Oxford University Press.

9

Realism and instrumentalism about the wave function: how should we choose?

MAURO DORATO AND FEDERICO LAUDISA

9.1 Introduction

It is not exaggeration to claim that one of the major divides in the foundations of non-relativistic quantum mechanics derives from the way physicists and philosophers understand the status of the wave function. On the instrumentalist side of the camp, the wave function is regarded as a mere instrument to calculate probabilities that have been established by previous measurement outcomes.[1] On the other "realistic" camp, the wave function is regarded as a new *physical* entity or a *physical* field of some sort. While both sides agree about the existence of quantum "particles" (the so-called theoretical entities), and therefore reject the radical agnosticism about them preached by van Fraassen (1980), the various "realistic" (and consequently, instrumentalist) philosophies of quantum mechanics are typically formulated in different, logically independent ways, so their implications need to be further investigated.

For instance, on the one hand it seems plausible to claim that a realistic stance about the wave function is not the only way to defend "realism" about quantum theory. One can support a "flash" or a "density-of-stuff" ontology (two variants of GRW), or an ontology of particles with well-defined positions (as in Bohmian mechanics), as *primitive* ontologies for observer-independent formulations of quantum mechanics (Allori, Goldstein, Tumulka and Zanghì, 2008). "Primitive ontologies", as here are understood, are not only a fundamental ground for other ontological posits, but also entail a commitment to something concretely existing in spacetime (see also Allori, 2013). On the other hand, however, it is still debated

---

[1] Among representatives of this form of instrumentalist, one can cite, among many others, Bohr (1972 2006) and Rovelli (1996). For Bohr's antirealism about quantum theory (and realism about quantum entities) see Faye (1991). For Rovelli's analogous stance, we refer the reader to Dorato (2013). For an exposition of Rovelli's relational interpretation, see Laudisa and Rovelli (2013). Here, we don't worry about the tenability of the distinction between entity realism and theory realism.

whether such primitive ontologies can be autonomous from some form of realism about the wave function (Albert, 1996).

In order to discuss this problem, we begin with a preliminary clarification of the meaning of "realism" and "instrumentalism" in physics, which are often subject to ideological and abstract discussions that often have little to do with the practice of physics (Section 9.2). In the following sections we present the various forms that a realism about the wave function can take; namely, in Section 9.3 we assess configuration-space realism (Albert, 1996), or wave function-space realism (North, 2013), a form of realism that might be backed up by Psillos' (2011) realist and "literalist" attitude toward the abstract models postulated by physics. In Section 9.4 we discuss what we call Ψ-nomological realism – or realism about the guiding law of Bohmian mechanics – as a consequence of a more general primitivism about physical laws defended in Maudlin (2007).[2] Considering the wave function of the Universe as a *nomological object* is a way of defending this position (Goldstein and Zanghì, 2013, p. 96). In Section 9.5 we present a form of *indirect* wave-function realism, according to which the wave function indirectly refers to real physical properties, for instance in virtue of the eigenvalue–eigenvector link: "the wave function doesn't exist on its own, but it corresponds to a property possessed by the system of all the particles in the Universe" (Monton, 2006, p. 779). The recent dispositionalism about quantum properties seems a way to formulate this position (Dorato, 2007b; Dorato and Esfeld, 2010; Esfeld, Lazarovici, Hubert and Dürr, 2013). In Section 9.6 we evaluate a much debated wave-function realism, according to which the quantum state (as described by the wave function) is independent of the knowledge of the observer, so that it is more than mere "information" that observers have about the system (Pusey, Barrett and Rudolph, 2012).

A natural question is which of these various ways of formulating realism about the wave function (RWF for short) is more plausible, in the hypothesis that they are all independent of each other. Providing an answer to this question (and therefore to the problem whether instrumentalism about the wave function is not the most reasonable position to take) is the main target of our chapter.

9.2 Realism as a stance and its pluralistic consequences

Is it possible to discuss the ontological status of the wave function independently of a specific interpretation of quantum mechanics? In order to answer this question in the affirmative, some considerations on the realism/instrumentalism debate seem appropriate.[3] In our opinion, to be a realist about physical theories in general

---

[2] "I suggest to regard laws as fundamental entities in our ontology" (Maudlin, 2007, p. 18).
[3] For lack of space, here they will be have to be taken for granted.

is a *stance* (van Fraassen, 2002), that is, an attitude toward the aim of physical theorizing. This assumption entails that there is no *a priori* guarantee that such an aim will be accomplished in all cases or by all theories. Often, scientists and philosophers are able to tell – and history can teach us – when a realist approach to a given theory is justified or not. It then becomes not unreasonable to be instrumentalist about physical theory *x* and realist about theory *y*, according to the kind of evidence (and other epistemic virtues) that *x* and *y* can boast.[4] Even more radically, one can be instrumentalist about *different components* of the *same* physical theory: Lange (2002), for example, argues convincingly that one ought be realist about the electromagnetic field but antirealist about Faraday's lines of force.

If we adopt the above-mentioned anti-ideological and pragmatic attitude toward scientific realism in general, an evaluation of the *pros* and *cons* of the various kinds of wave-function realism does not *a priori* force us to take a stand in favor of or against a particular type of a primitive ontology for quantum theory. Our inquiry can be important for evaluating the different interpretations of quantum mechanics with respect to the status of the wave function; these interpretations and their mutual relations in fact cannot be represented exclusively in terms of logical implications between the above mentioned primitive ontologies (PO) and the different forms of wave-function realism (WFR). To exemplify, let us consider the two following possibilities.

(1) Let us suppose that the assumption of a primitive ontology requires some form of realism about the wave function as a necessary condition (PO → WFR). If this were the case, instrumentalists about the wave function could reject primitive ontologies via a simple *modus tollens*. In this first alternative, the question of inquiring into the reality of the wave function *per se* assumes a particular importance, but Bohmian mechanics turns out to be a counter-example to the claim that we *need* to treat the wave function as a robustly real entity in its own right in order to be justified in assuming a primitive ontology.

(2) The converse implication (WFR → PO) amounts to assuming that an attribution of some type of ontological status to the wave function presupposes a primitive ontology of a sort as its necessary condition. Again, there seems to be a counter-example to the complete generality of such an implication: in the configuration space realism defended by Albert, elevating the configuration space in which the wave function lives to the status of ultimate reality need not imply the requirement of a primitive ontology of entities in spacetime as the primary objects the theory is about, since the theory in the Albert sense is *primarily* about the configuration space itself.

---

[4] For this viewpoint, see Dorato (2007a).

In both cases, anyway, establishing in what sense the wave function can be an "element of reality" will have interesting implications for the kind of primitive ontology that is more plausible to adopt. Since these brief remarks should suffice to justify the focus of our chapter on wave-function realism, we can proceed to discuss the various options at stake.

9.3 Realism about configuration space

John Bell once wrote about quantum mechanics: "no one can understand this theory until he is willing to think of ψ as a real objective field rather than just a "probability amplitude". Even though it propagates not in 3-space but in $3N$-space" (Bell, 1987, p. 128). David Albert takes inspiration from this passage, as in his view (Albert, 1996, 2013), the wave function is regarded as a physical *field*. It is often presupposed that since any physical field is an assignment of values to a space, the space on which the field sits must be regarded as real. As is well known, however, the wave function can be an assignment of physical magnitudes (positions, for Bohmian mechanics) to every point of 4D spacetime only if we have a one-particle system. As soon as N particles are considered, the wave function lives in a $3N$ configuration space: "The sorts of physical objects that wave functions *are,* on this way of thinking, are (plainly) *fields* – which is to say that they are the sorts of objects whose states one specifies by specifying the values of some set of numbers at every point in the space where they live, the sorts of objects whose states one specifies (in *this* case) by specifying the values of *two* numbers (one of which is usually referred to as an *amplitude,* and the other as a *phase*) at every point in the Universe's so-called *configuration* space" (Albert, 1996, p. 277).

One widely recognized, first problem with this view is how one can recover tables and chairs occupying a 4-dimensional spacetime (namely POs in the sense of Allori *et al.*, 2008) from a $3N$-dimensional configuration space. Using magic words like "emergence" is not going to help: until a convincing explanatory sketch of such an emergence is available, we submit that one has no reason whatsoever to take configuration space realism seriously.

It could be replied that while science is a sophistication of common sense, it is often capable of reaching conclusions that cast radical doubts on important *components* of common sense. Our first argument against this reply is that the stress in the previous sentence is on "components". Notice the difference from past episodes in the history of science. For example, when natural philosophers discovered that the Earth is not stationary, they had to explain how it could be in motion without us noticing it. The reconciliation of the scientific worldview with the world of our senses was achieved via the introduction of the notion of inertia. An analogous explanation was achieved of our natural belief in the worldwide nature of the present moment, which was later superseded by Einstein's postulation of the relativity of simultaneity. In fact, one can explain why we tend to believe that the present moment has cosmic extension in terms of the speed of light and the finite duration needed by our brain to process temporally successive light signals (Dorato, 2011).

However, in the case of configuration space realism, it is the whole worldview of common sense that is regarded as "misleading", and since science relies on observations and therefore on common sense, the consequence that all our observations are radically illusory cannot be accepted.

It must be admitted that quantum mechanics requires anyway an important sacrifice of elements of the manifest image; but in this regard even Everettian quantum mechanics is in better shape with respect to the task of explaining the emergence of our spacetime from configuration space, insofar as it can explain with the help of decoherence why the local observer cannot perceive any interference with the other worlds. In other words, if believing that the wave function is a physical entity (a field) implies configuration realism, it could be argued by *modus tollens* that the wave function is not physically real. Since the abstract or concrete ontological status of the wave function will be discussed in later sections, let us assume that the wave function might be neither concrete nor abstract, and yet a wholly new physical entity (Maudlin, 2013). After all, why should we assume that something is physical only if it is in 4D spacetime?

Two remarks are sufficient to create troubles for this assumption. First, the case of strings, which live in compactified dimensions, is different from that of a $3N$-dimensional field. Strings still live in spacetime, even though the latter is conceived as being ten-dimensional or even 26-dimensional. The fact is that the extra dimensions are too small to be "seen" (compactification). The second difficulty is given by the fact that the problems that afflict configuration space realism also arise in the case of a multidimensional ($3N$) physical field. How can we recover a four-dimensional field (say the electromagnetic field) from the former? Until an explanatory sketch is provided, there is no reason to reify the wave function by requiring that the mathematical space needed to define it is the real stuff the Universe is made of. As Maudlin notes (2013, p. 152), mathematical representations of physical phenomena are not a clear guide to ontology, since they often do not guarantee even isomorphic relations between themselves and the latter. Furthermore, for obvious algorithmic reasons they must greatly simplify and idealize the target they are a vehicle for, and so they are not necessarily similar to what they are supposed to denote.

A different form of realism about the wave function has been defended by North (2013), who distinguishes between *configuration*-space realism and *wave*-function realism, a kind of ontic structural realism about the latter. Here we can afford to be brief about her interesting proposal, since she assumes rather than arguing that the wave function "directly represents or governs" parts of the ontology of quantum mechanics (North, 2013, p. 185). Her main stance is a form of epistemic primitivism about laws, since she claims that dynamical laws of a theory are our main guides to infer what exists according to the theory, and what exists at the fundamental level is the structure that is needed to formulate the laws. What is missing in her semi-transcendental approach is the validation of the claim that there is only *one* mathematical way to formulate the dynamic laws, a step that is necessary to claim uniqueness also for the inferred physical structure. In fact she denies any guiding role to Hilbert space (North, 2013, p. 191) and she does not even mention other more algebraic and abstract formulations of quantum mechanics; but it is not wholly clear on the basis of which criterion this selection is suggested: whether a state space has too little or superfluous structure typically depends on the problem at hand. And we want to add that, not by chance, such "uniqueness questions" are a typical problem for any form of ontic structural realism, since it is highly difficult to prove that the same dynamical laws cannot be formulated by presupposing a different mathematical structure.[5]

9.4 The wave function as a nomological entity

In Goldstein and Zanghì (2013), the wave function is defined as a "nomological entity", the primitive ontology being constituted by the positions of particles in spacetime, or by the actual positions of the particle $\mathbf{Q} = (Q_1, Q_2, Q_3, Q_4, \ldots, Q_n)$ in configuration space. Since the two authors are not terribly clear about what we should mean by "nomological entity" (are physical laws entities?), it is important to defend their position as best we can in order to overcome initial resistances of philosophers to the infelicitous choice of the term.

First, evidence for a robust ontic status of Ψ is suggested by its role in Bohm's "guiding" equation: the velocity of any of the N particles is a function via Ψ of the positions of all the other particles. Second, according to Goldstein and Zanghì, the real nomological entity is properly speaking only the wave function of the Universe, since the Universe is "the only genuine Bohmian system" (Goldstein and Zanghì, 2013, p. 94), the wave function of a subsystem being only definable in terms of the wave function of the Universe and the whole set of configuration of all the particles. Given the fundamental non-locality of the theory, this is only to be expected, even though for all practical applications what one deals with in Bohmian

---

[5] This problem is no less acute in spacetime theories, where general relativity can be formulated in a variety of different mathematical formalisms (that of Riemannian differentiable manifolds, Einstein algebras, twistors, non-commutative geometry and so on).

mechanics are subsystems. Since this presupposes the possibility of attributing a wave function to the Universe, it seems legitimate to ask whether this move is legitimate,[6] given the present lack of a quantum theory of gravity, or even of a well-worked out relativistic extension of Bohmian mechanics. Despite the fact that at the moment the attribution of a wave function to the Universe is rather speculative, or even devoid of any clear empirical meaning, for the sake of the argument we will assume without further ado that our ontological quest is limited to a Newtonian, non-relativistic spacetime, which possesses a privileged foliation.

Given these two clarifications, the real question (our third point) is of course how to understand the ontological status of the wave function as a nomological entity. It will not do to invoke vague metaphors like the wave function "choreographs" or "governs" the motion of the particles, since laws strictly speaking do not govern like kings: if they literally governed, they would have to be "external" (to continue the metaphor) to what they govern. But if they are external, how can they affect physical entities in the sense in which Ψ must "guide" the motion of particles? This governing view seems a remnant of a theological, prescriptive rather than descriptive conception of laws, motivated by the hypothesis that a Creator imposes his own will to Nature, its creature (Dorato, 2005, chapter 1).

Abandoning, as it is fair to do, the literal interpretation of the term "governing", there is still an important question that needs to be raised *apropos* of the wave function regarded as a "nomological entity": are nomic entities in general external or internal to the entities and the properties that instantiate them? This issue is important in order to clarify the property-first view of laws vs. nomic primitivism, and therefore how we should understand Zanghì and Goldstein's view of Ψ as nomological. Moreover, it leads to specifying three different senses of "primitive": the first refers to the primitive ontology of spacetime-located entities the theory is about, the second refers to the *conceptually* non-reductive character of the notion of lawhood which the primitivism about laws is grounded upon and, finally, the sense in which such "special" nomological entities as the wave functions are *ontologically* primitive.

Non-metaphorically, the term "external", when referred to laws, typically means "independent or non-supervenient upon the entities and the properties they relate", while "internal" is therefore equivalent to "dependent on those entities and properties". As Psillos put it, "external" means that the laws can vary while the properties that they instantiate do not change (Psillos, 2006, p. 18) and this implies a sort of *quidditism*. This is the view that there are properties P whose identity is independent of, and can be detached from, their nomic or causal role R, so that it is

---

[6] In the Rovelli relational interpretation, for instance, such an attribution makes no sense (Laudisa and Rovelli, 2013; Dorato, 2013).

not essential to a property that it plays a given nomic role. It should be admitted that quidditism, exactly like heacceitism,[7] cannot be ruled out *a priori*. However, it is certainly difficult to accept the view that the property P that electrons possess of being negatively charged – which entails the nomic role R to attract positively charged bodies – could be detached from R in such a way that P would remain the *same* even if governed by a different law (and therefore be characterized by a different R). Be that as it may, the other horn of the dilemma (laws as internal to properties) implies a "property-first" view on laws, and therefore the idea that laws supervene on properties and relations of entities and cannot be ontically primitive, let alone "govern" their instances.

Leaving aside the metaphysical complications of quidditism, for us it is important to note that the choice between these two alternatives ("externalism" or "internalism" about laws) does not force one to be antirealist about laws,[8] a position that would rule out the possibility that Ψ is a nomological entity in the sense of Goldstein and Zanghì. Nevertheless, in the remainder of this section we will assume that their view is committed to primitivism (or the non-supervenience of laws on properties) for essentially two reasons. First, the "internal", second alternative pushes toward nomic antirealism, since the properties or the powers of entities could exhaust all the roles played by laws (Mumford, 2004). Second, the property-first view of laws implied by internalism will be discussed in the next section.

Suppose then that wave-function realism is committed to some sort of *ontic* primitivism about laws in the sense of Maudlin (2007). The problem is that once one abandons the safe ground offered by the *conceptual* priority (or irreducibility) of nomic concepts in the sense of Carroll (1994) (the second sense of primitive above), it is not clear what *ontic* primitivism amounts to. On the one hand, we cannot assume without further arguments that conceptual priority entails ontic priority, since the concept of *scientific law* might be irreducible to other related concepts (causation, counterfactuals, regularity, etc.), without implying any sort of ontic primitivism about laws of nature.[9] On the other hand, if one does not want to beg the question against primitivism, it must be admitted that there is a sense in which ontic nomic primitivism cannot be further understood, precisely because the notion of law is regarded as un-analyzable.

However, this irreducibility might be regarded as a serious deficiency of this position for at least two reasons.

---

[7] Quidditism is the view that properties have an intrinsic nature that is not exhausted by their causal role. Heacceitism is the view that entities have an intrinsic identity that is not reducible to their properties. So two individuals can be different even if they are qualitatively identical.

[8] One could claim that laws exist but that they are just relations between entities that are primary and more fundamental.

[9] The concept of knowledge might be irreducible to justified true belief, and yet knowledge is not ontically primitive.

(1) It is true that we must start from somewhere, in mathematics as well as in philosophy: it is the explanatory consequence of taking a notion *A* as primitive that justifies the choice of *A*. However, mathematics relies on axioms, which give an implicit definition of the axiomatized notions. In philosophy, on the contrary, when we do not understand a notion (in this case "laws of nature regarded as existent"), we seem to be in a different and more difficult predicament. When a concept *A* is more obscure than a concept *B*, and we declare *A* "primitive" – laws seem to be less intuitively understood than properties – we run the risk of wanting to solve a philosophical problem without even trying.

This difficulty, however, can be solved: after all, intuitions about what is obscure may vary. Let us then agree that a fair reading of "ontically primitive" with respect to laws might mean, simply, that *there are mind-independent nomic facts* that are the supervenience basis for the existence of those properties, dispositions, causal facts and the like that (according to the primitivists) are mistakenly regarded as the truth-makers of the propositions that express the "laws of science".

(2) This formulation brings with itself the second difficulty. Since these (approximately true) propositions regarded as truth-bearers in physics are typically differential equations, for the primitivist about laws the existence of *nomic* (physically necessary) facts must be contrasted with the existence of merely *contingent* facts, typically lying in hypersurfaces of simultaneity, and specifying the initial or boundary conditions to which the equations are applied. But how can the primitivist distinguish between the modally loaded, nomic facts, and the contingent facts, if both are *facts*? Clearly, ontic primitivists about laws cannot ground the distinction between nomic and contingent truths on the existence of physically possible worlds, lest law loses its primitivity. The same applies to purely conceptual primitivism. Furthermore, note that in this rendering, ontic primitivism has to be realist about the existence of facts and must regard them as *concrete* entities, being in any case distinct from *abstract* propositions.

If we apply these two objections to our problem, the difficulty should be obvious: claiming that the wave function is an entity because the laws in which it appears exist in a primitive sense is not convincing, because a physical hypothesis is made to depend on a highly controversial metaphysical hypothesis.

9.5 The property-first view of the wave function: dispositionalism

We have seen that according to *primitivism* about, say, the guiding law of Bohmian mechanics, there is in the quantum world a global, nomic fact instantiated by the world in question that determines the temporal development of an initial, contingent configuration of particles belonging to a hypersurface of simultaneity. According to the property-first view, it is instead the initial configuration of point-particles in a background spacetime that, by instantiating a plurality of properties, fixes the temporal evolution of whatever exists in the initial configuration. In the literature, such properties have often been regarded as dispositions, so that the subsequent behavior of the initial configuration of particles is given by their manifestations.

Just to exemplify, in the case of flashy GRW, the disposition of non-massless entities to localize in a flash, or in a region of spacetime in the mass-density reading of GRW, is a spontaneous and an irreversible process. The flash or a certain localized mass density in spacetime is the manifestation of the disposition in question. In the case of Bohmian mechanics, each particle has a spontaneous disposition to influence the velocity of the i-th particle in a non-local way, and the velocity of that particle is the manifestation of the global disposition carried by the whole configuration of particles (Esfeld, Lazarovici, Hubert and Dürr, 2013). Thus, on Bohmian mechanics the configuration of *all* particles at a given time t instantiates a *dispositional property* that manifests itself in the velocity of each particle at t; the universal wave function at t represents that property, so that the latter is ontologically primary and the wave function refers to such a property.[10]

The difference with GRW's two primitive ontologies is that the dispositions in the latter are really probabilistic propensities (GRW is irreducibly indeterministic), while in Bohmian mechanics they are sure-fire dispositions. But in both cases (deterministic and indeterministic), the introduction of dispositional properties has the advantage of avoiding a reification of the configuration space.

However, in both cases there are two difficulties that all these property-first views share: (i) quantum dispositions are spontaneous but in standard situations classical, typical dispositions need a stimulus (a stone breaking a window pane, with the ensuing manifestation of fragility being the broken glass); (ii) physical laws, referring to or representing dispositions, are, unlike dispositions, *time-symmetric*.

The first difficulty (i) depends on how one defines dispositions, namely in a more liberal or in a less liberal way, so as to encompass also spontaneous manifestations. We think liberalism about this issue can be justified, in order not to beg the question against dispositionalism. Any mass has a spontaneous disposition to move inertially, even though the disposition to resist acceleration is manifested only in the presence of a force (the stimulus). Likewise a radioactive material has a spontaneous disposition (a propensity) to decay, even though the decay can be accelerated by bombarding the nucleus in an opportune way.

---

[10] For a more detailed description of this view, see Dorato and Esfeld (2014).

In the two GRW cases, the second difficulty (ii) could be more easily accommodated by treating the new non-linear equations introduced by the dynamical reduction models as time-asymmetric *laws*, namely nomic irreversibilities that explain or ground the less fundamental arrows of time (as suggested by Albert, 2000). In the Bohmian case, the two main laws are time-symmetric, but one can hold that the irreversible dispositions to influence the velocity of each particle correlate to the arrow of becoming the successive occurrence of events given by the manifestation of the dispositions. In this way, only one of the two directions of time is the one in which the world unfolds, so that the temporal symmetrical feature would involve only the laws of *science* and not the laws of *nature*, which would take part in a universal process of becoming. Such a process can be regarded as either primitive (Maudlin, 2007, chapter 3), or explained by the manifestation of the various dispositions making true the laws of science (for the distinction between laws of science and laws of nature, see Weinert, 1995).[11]

9.6 The PBR theorem

As far as the controversy over the nature of the wave function is concerned, a new twist to the debate was provided by the so-called PBR theorem (Pusey, Barrett and Rudolph, 2012). According to a natural reading of this result, assuming the wave function of a quantum system S as a mere catalogue of the information available about S implies predictions that contradict those of quantum mechanics. As a consequence – we might argue – the idea that a quantum state is not just information about an entity but is a real entity *itself* should be taken seriously on physical and mathematical grounds. As a matter of fact, neither the general framework of the theorem nor the specific assumptions under which it is proved are innocuous, but before attempting an assessment let us recapitulate the result. The main hypothesis on the background is that "a system has a 'real physical state' – not necessarily completely described by quantum theory, but objective and independent of the observer" (Pusey, Barrett and Rudolph, 2012, p. 475). That such a state might be not completely characterized by quantum theory implies that a wave function ψ for a system S is taken to represent a *preparation* of the system itself: ψ fixes the "real" state λ non-uniquely but rather according to a probability distribution $\mu_\psi(\lambda)$. In the PBR approach – inspired by the terminology introduced in Harrigan and Spekkens (2010) – given two wave functions ψ and ϕ, we have two possible alternatives: either the probability distribution $\mu_\psi(\lambda)$ and $\mu_\phi(\lambda)$ do overlap or they do not. In the former case, there are values of the distributions that might be assigned to both ψ and ϕ, something that testifies to an uncertainty on what the "real" state associated

---

[11] This point has been initially suggested in Dorato and Esfeld (2014).

with either ψ or ϕ might be; in the latter case, the non-overlapping testifies to the lack of uncertainty: "informally, every detail of the quantum state is 'written into' the real physical state of affairs" (PBR, 2012, p. 476; Harrigan and Spekkens, 2010 speak of an *epistemic* view in the former case and of an *ontic* view in the latter). Under the additional assumption that independently prepared systems have independent physical states,[12] PBR prove that, given two distinct quantum states ψ and ϕ, the overlapping of the respective $\mu_\psi(\lambda)$ and $\mu_\phi(\lambda)$ implies a contradiction with the statistical predictions of quantum mechanics.

The PBR theorem is the n-th result of a long chain of *no-go theorems*, namely results that in principle should clarify the fundamental structure of the theory, by pointing out the boundaries that the theory itself is supposed not to violate when satisfying a class of basic constraints. Even leaving aside the general significance of the no-go strategy in the foundations of physics (Laudisa, 2014), there are several critical points that need be emphasized. The first is the most obvious but, nevertheless, the most urgent: one may ask what is the meaning of the assumption according to which "a system has a real physical state" when we lack a clear understanding of what it *means both* for the wave function ψ *and* the "real" state λ "to be real".[13] If it means that it is more than mere information, we still haven't been told much, that is, we have not been told what it is and what its properties are. As a consequence, the lack of a clear notion of what it takes for states like ψ or λ to be "real" implies that it is also completely unclear what it means that we *cannot* interpret ψ as mere information. It might be argued that, when we have ontic models, the quantum states *supervene* on "real" states, namely no change in quantum states without change in real states. Does this "supervenience" talk, however, help in understanding what it means to be "real", in the absence of an ontologically clear formulation of quantum mechanics itself? In some sense, both ψ- and λ-sort of states are supposed to carry with themselves an ontological stock that they in fact are unable to justify. For consider even the case of classical mechanics, that in the PBR approach is taken into account in order to explain the epistemic-versus-ontic view of states. In the Newtonian dynamics of a single point particle in one dimension, the description consists of specifying a point in the relevant phase space, namely a pair $\langle x(t_0),$ $p(t_0)\rangle$ at some given initial time t_0, where $x(t_0)$ is a position value and $p(t_0)$ is a momentum value at t_0: under the ideal assumption that we know all the forces at work, we can determine any pair $\langle x(t), p(t)\rangle$ at any given time t by using either the Newtonian or the Hamiltonian formulation of the dynamical laws. Now, it seems very natural at first sight to make sense of a pair like $\langle x(t), p(t)\rangle$ by stating that it is

---

[12] That this assumption is indeed necessary is proved in Lewis, Jennings, Barrett and Rudolph (2012); see also Schlosshauer and Fine (2012).

[13] That this is a problem can be seen also if we realize that (as PBR themselves remark) an instrumentalist is allowed to ignore the result of the PBR theorem, unlike the case, for instance, of the measurement problem, which is at least partly a problem *also* for the instrumentalist.

a clear instance of an "ontic state", namely of "a state of reality". Since, however, that pair is in fact simply a point in an abstract, multi-dimensional configuration space, it must be noted that in order for such a "classical" pair to be a "state of reality", a rather heavy assumption must be accepted, namely that what is "really" real is not our three-dimensional experience but rather the manifold with an astronomical number of dimensions whose points are all the possible ontic states determined by the classical dynamical laws. Therefore this reference to the supposedly more familiar case of classical mechanics on phase space, far from serving the purpose of PBR of enlightening the meaning of what a "real" state should be, shows that there is a big gap between certain mathematical structures of the state space on one side and the realm of "real" states (whatever they might be).

9.7 Conclusion

After presenting and discussing the main options available on the status of the wave function in the foundations of quantum mechanics, it may be worthwhile trying to recapitulate the general framework and possibly to draw some connections among the different approaches. As far as the latter are concerned, they all display a significant metaphysical flavor inasmuch as they all adopt robust metaphysical assumptions concerning their respective target entities – namely the configuration space, the laws of nature and natural properties; as to their respective plausibility, they can be evaluated on how well they fare with respect to our intuition and common sense on one side and to the role they might play in the foundations of quantum physics on the other.

If we start with the configuration space realism, we might argue that the weight of the usual objection – according to which the reality at the level of the configuration space would be hard to reconcile with the reality at the level of our ordinary, three-dimensional experience of physical systems and processes – is debatable. While common sense certainly sides with three-dimensional experience, it is also true that the extent to which common sense is really "common" might be controversial. Since common sense is a vague notion, certain tenets of common sense might be subjective, as different people might have different views on what counts as important in "common sense". Furthermore, the abstractness of the configuration space might also not be an unsurmountable problem in itself (one can adopt, for instance, the motivations defended by Psillos, 2011).

The above-mentioned remark by Maudlin, however (namely that mathematics is not often a safe guide to ontology), should be taken into due account (examples abound: recall either the algebraic formulation of quantum field theory or the quantum logic program). Moreover, a serious problem with the configuration space realism can be the attempt to supplement the thesis that the only reality is configuration space with the highly controversial assumption of Humean supervenience (HS for short; see Loewer, 1996; Darby, 2012): the latter assumption would be supposed to dispense with the problem of what is the sense in which a wave function "lives" and plays its statistical and dynamical role in a configuration space, since according to a configuration-space+HS stance only the Lewisian mosaic of facts would exist and nothing else (putting aside the problem of what is a "fact" in configuration space: a point $x = (q_1, q_2, \ldots)$, the fact that a point x has coordinates $(q_1, q_2, \ldots)$, or what?).

That we need not worry if, for good reasons, we are led to include abstract entities in the inventory of the world seems to apply also to the nomological approach to the wave function. If we are prepared to contemplate a $3N$-dimensional manifold as the ultimate physical reality, no less prepared should we be to contemplate the existence of nomological "entities", whatever they might be. But this puts Goldstein and Zanghì nomological-entity realism on the same footing as Albert configuration space realism. Both bet on abstract entities. Moreover, on the face of possible objections – some of which have been mentioned before – to the primitivist view of laws, on which the nomological approach seems to be most naturally grounded, it might be remarked anyway that the primitivist view seems to have at least an advantage over the property-first view in terms of both conceptual and metaphysical economy. In fact, the dispositional reading of the property-first view applied to quantum mechanics seems to imply a worldview in which the quantity of dispositional properties amounts to the quantity of particles that are supposed to display a certain behavior under certain conditions: instead of giving up laws and having an astronomical number of particles, each with its bundle of dispositional properties, would it not be "easier" to have a restricted number of laws that account for the seemingly dispositional behavior of the particles? If, on the other and, staying closer to the spirit of Esfeld *et al.* (2013), the many dispositions of the single particle in question really amounted to a unique disposition of the whole configuration space described by the wave function, then primitivism and global dispositionalism would seem to converge and the distinction between properties-first and laws-first might be purely verbal and lose some of its importance. The global nomic fact that, according to primitivism, is instantiated by the quantum world would correspond to the global disposition characterizing the configuration space in the sense of Esfeld *et al.*[14]

Of course, any nominalistic philosopher of human sympathies would be inclined to reject any form of commitment to abstract entities like laws or configuration spaces, and embrace wave function instrumentalism *sic et simpliciter*. And also

[14] For this claim, see Dorato and Esfeld (2014).

this latter position is certainly not incompatible with what we know about the physics of quantum theory. In a word, the most plausible moral to be drawn at this point is that the metaphysics of Ψ is radically undetermined by quantum physics and even by the sort of primitive ontology one adopts, a conclusion which need not hold for all metaphysical claims in their relation with physical theories.

References

Albert, D. Z. (1996) Elementary quantum metaphysics, in J. T. Cushing, A. Fine and S. Goldstein (eds.), *Bohmian Mechanics and Quantum Theory: an Appraisal.* Dordrecht: Kluwer, pp. 277–284.

Albert, D. Z. (2000) *Time and Chance*, Harvard: Harvard University Press.

Albert D. Z. (2013) Wave function realism, in A. Ney and D. Albert (eds.), *The Wave Function*, Oxford: Oxford University Press, pp. 52–57.

Allori, V. (2013) Primitive ontology and the structure of fundamental physical theories, in A. Ney and D. Albert (eds.), *The Wave Function*, Oxford: Oxford University Press, pp. 58–75.

Allori, V., Goldstein, S., Tumulka, R. and Zanghì, N. (2008) On the common structure of Bohmian mechanics and the Ghirardi–Rimini–Weber theory, *British Journal for the Philosophy of Science*, **59**: 353–389.

Bell, J. S. (1987) *Speakable and Unspeakable in Quantum Mechanics*, Cambridge: Cambridge University Press.

Bohr, N. (1972–2006) *Collected Works*, Vol. 1–12, Amsterdam: Elsevier.

Carroll, J. (1994) *Laws of Nature*, Cambridge: Cambridge University Press.

Darby, G. (2012) Relational holism and Humean supervenience, *British Journal for the Philosophy of Science,* **63** (4): 773–788.

Dorato, M. (2005) *The Software of the Universe. An Introduction to the History and Philosophy of Laws of Nature,* Ashgate: Gower-Lund Humphries.

Dorato, M. (2007a) *Cosa c'entra l'anima con gli atomi?*, Roma: Laterza.

Dorato, M. (2007b) Dispositions, relational properties and the quantum world, in M. Kistler and B. Gnassonou (eds.), *Dispositions and Causal Powers*, Farnham: Ashgate, pp. 249–270.

Dorato, M. (2011) The Alexandroff present and Minkowski spacetime: why it cannot do what it has been asked to do, in D. Dieks, W. Gonzales, S. Hartmann, T. Ubel and M. Weber (eds.), *Explanation Prediction and Confirmation. New Trends and Old Ones Considered.* Amsterdam: Springer, pp. 379–394.

Dorato, M. (2013) Rovelli's relational quantum mechanics, monism and quantum becoming, arxiv.org/abs/1309.0132.

Dorato, M. and Esfeld, M. (2010) GRW as an ontology of dispositions, *Studies in History and Philosophy of Modern Physics*, **41** (1): 41–49.

Dorato, M. and Esfeld, M. (2014) The metaphysics of laws: dispositionalism vs. primitivism, forthcoming.

Esfeld, M., Lazarovici, D., Hubert, M. and Dürr, D. (2013) The ontology of Bohmian mechanics. Forthcoming in *British Journal for the Philosophy of Science* **64**. philsci-archive.pitt.edu/9381/, doi:10.1093/bjps/axt019.

Faye, J. (1991) *Niels Bohr: His Heritage and Legacy. An Antirealist View of Quantum Mechanics*, Dordrecht: Kluwer Academic Publisher.

Goldstein, S. and Zanghì, N. (2013) Reality and the role of the wave function in quantum theory", in A. Ney and D. Albert (eds.), *The Wave Function*, Oxford: Oxford University Press, pp. 96–109.

Harrigan, N. and Spekkens, R. W. (2010) Einstein, incompleteness, and the epistemic view of quantum states, *Foundation of Physics* **40**, 125–157.

Lange, M. (2002) *An Introduction to the Philosophy of Physics*, Oxford: Blackwell.

Laudisa, F. and Rovelli, C. Relational quantum mechanics, in E. N. Zaltia (ed.), *The Stanford Encyclopedia of Philosophy*, plato.stanford.edu/archives/sum2013/entries/qm-relational/.

Laudisa, F. (2014), Against the 'no-go' philosophy of quantum mechanics, forthcoming in the *European Journal for the Philosophy of Science*.

Lewis, P. G., Jennings, D., Barrett, J. Rudolph, T. (2012), Distinct quantum states can be compatible with a single state of reality, *Physical Review Letters*, **109**, 150404.

Loewer, B. (1996) Humean supervenience, *Philosophical Topics*, **24**, 101–127.

Maudlin, T. (2007) *The Metaphysics within Physics*, Oxford: Oxford University Press.

Maudlin, T. (2013) The nature of the quantum state, in A. Ney and D. Albert (eds.), *The Wave Function*, Oxford: Oxford University Press, pp. 126–154.

Monton, B. (2006) Quantum mechanics and 3-N dimensional space, *Philosophy of Science*, **75** (5), 778–789.

Mumford, S. (2004) *Laws in Nature*, London: Routledge.

North, J. (2013) The structure of a quantum world, in A. Ney and D. Albert (eds.), *The Wave Function*, Oxford: Oxford University Press, pp. 184–202.

Psillos, S. (2006) Review symposium, looking for laws, *Metascience*, **15**, 452–464.

Psillos, S. (2011) Living with the abstract, *Synthese*, **180** (1), 3–17.

Pusey, M., Barrett, J. and Rudolph, T. (2012), On the reality of the quantum state, *Nature Physics*, **8**, 475–478.

Rovelli, C. (1996) Relational quantum mechanics, *Int. J. Th. Phys.* **35**, 1637.

Schlosshauer, M. and Fine, A. (2012) Implications of the Pusey–Barrett–Rudolph quantum no-go theorem, *Physical Review Letters*, **108**, 260404.

van Fraassen, B. (1980) *The Scientific Image*, Oxford: Oxford University Press.

van Fraassen, B. (2002) *The Empirical Stance*, New Haven and London: Yale University Press.

Weinert, F. (1995) *Laws of Nature. Essays on the Scientific, Philosophical and Historical Dimensions*, Berlin: de Gruyter.

10

Protective measurement and the PBR theorem

GUY HETZRONI AND DANIEL ROHRLICH

Protective measurements illustrate how Yakir Aharonov's fundamental insights into quantum theory yield new experimental paradigms that allow us to test quantum mechanics in ways that were not possible before. As for quantum theory itself, protective measurements demonstrate that a quantum state describes a single system, not only an ensemble of systems, and reveal a rich ontology in the quantum state of a single system. We discuss in what sense protective measurements anticipate the theorem of Pusey, Barrett, and Rudolph (PBR), stating that, if quantum predictions are correct, then two distinct quantum states cannot represent the same physical reality.

10.1 Introduction

Although protective measurements [1, 2] are a new tool for quantum theory and experiment, they have yet to find their way into the laboratory; also theorists have not put them to best use, beyond a 1993 paper by Anandan on "Protective measurement and quantum reality" [3]. In Section 10.2, we point out that protective measurements offer new experimental tests of quantum mechanics, and we review recent experiments attempting to measure quantum wave functions. In Section 10.3, we present the Pusey–Barrett–Rudolph (PBR) theorem and discuss their conclusion that the quantum state represents physical reality, and in Section 10.4, we discuss in what sense protective measurements anticipate this conclusion.

10.2 Protective measurement: implications for experiment and theory

In 1926, Schrödinger postulated his equation for "material waves" in analogy with light waves: paths of material particles – which obey the principle of least action – are an approximation to material waves, just as rays of light – which obey the principle of least time – are an approximation to light waves [4]. But Born soon discarded "the physical pictures of Schrödinger" [5] and gave the "material wave" $\Psi(\mathbf{x}, t)$ a new interpretation: $|\Psi(\mathbf{x}, t)|^2$ is the probability density to find a particle at $\mathbf{x}$ at time t. Even Schrödinger was obliged to accept Born's interpretation. But Born's interpretation limits the correspondence between quantum theory and experiment, in the following sense: for a *single* particle, $\Psi(\mathbf{x}, t)$ seems *not* to be measurable; to measure a probability density, we need to prepare $\Psi(\mathbf{x}, t)$ on an ensemble. Thus, part of what quantum theory describes – the wave function $\Psi(\mathbf{x}, t)$ of a single particle – does not correspond to anything experiments can measure. The paradigm of protective measurement [1, 2, 6], by contrast, makes the correspondence explicit: experiments *can* measure the wave function of a single particle! Protective measurements make it possible to measure the expectation value of any operator A in any state $|\Psi\rangle$ using a single system prepared in the state $|\Psi\rangle$, and thus to reconstruct $|\Psi\rangle$. An ensemble of identical systems in the state $|\Psi\rangle$ is not necessary. By the same token, the method of protective measurement allows us to test quantum mechanics in ways that were never considered before, i.e. to verify expectation values measured on an isolated system.

A recent experiment of Lundeen et al. [7, 8] measured the transverse spatial wave function of a photon propagating as a plane wave. These authors do not mention "protective measurement" – they refer only to "weak measurement" – and their experiment differed from a protective measurement in two ways. First, the measurement was applied to an ensemble of photons rather than to a single photon; second, what they measured was not an expectation value but the weak value [9] of the projection operator $\Pi_x \equiv |x\rangle\langle x|$ onto a transverse position x:

$$\langle \Pi_x \rangle_w = \frac{\langle p|x\rangle\langle x|\Psi\rangle}{\langle p|\Psi\rangle} = \frac{\mathrm{e}^{-\mathrm{i}px/\hbar}\Psi(x)}{\langle p|\Psi\rangle}, \tag{10.1}$$

where $\Psi(x)$ is the (preselected) transverse wave function to be measured, $|p\rangle$ is a (postselected) transverse momentum eigenstate with momentum p, and the postselected momentum is $p = 0$. Then the weak value is proportional to $\Psi(x)$ and the initial wave function (both real and imaginary parts) is measured as a function of x.

Although the weak measurement of Lundeen et al. is not a protective measurement, protective measurements are a form of weak measurement [10]. If the pre- and postselected states $|\Psi_{\mathrm{in}}\rangle$, $|\Psi_{\mathrm{fin}}\rangle$ of a weak measurement of A are the same, the measured weak value $\langle A\rangle_w$ is the expectation value of A in the state $|\Psi_{\mathrm{in}}\rangle$:

$$\langle A \rangle_w = \frac{\langle \Psi|A|\Psi\rangle}{\langle \Psi|\Psi\rangle} = \langle A \rangle, \tag{10.2}$$

where $|\Psi_{\mathrm{fin}}\rangle = |\Psi\rangle = |\Psi_{\mathrm{in}}\rangle$. In a typical weak measurement, the pointer of a measuring device is coupled to an *ensemble* of systems pre- and postselected in the state $|\Psi\rangle$, and shifts as the expectation value $\langle A \rangle$ accumulates from all the systems in the ensemble. (Note that, while the postselection in many weak measurements is improbable, here the postselection is the most probable.) By contrast, a protective measurement is essentially a weak measurement repeated on the *same* system, and the pointer shifts as the expectation value $\langle A \rangle$ accumulates from the repeated measurement. Repeated post- and preselections insure the protection, and most-probable postselections insure the adiabaticity. In effect, a repeated measurement of A on a single system in the state $|\Psi\rangle$ yields the same result as single measurements of A on an ensemble of systems pre- and postselected in the state $|\Psi\rangle$; however, only the first kind of measurement – protective measurement – explicitly manifests the quantum state of a *single* system.

A more recent experiment by Stodolna et al. [11] maps the nodal structure of the $n = 30$ Rydberg level of hydrogen in a uniform electric field. (See also a related experiment by Cohen et al. [12], and measurement of molecular wave functions by Lüftner et al. [13]. These three papers, as well, do not mention protective measurements.) An electron excited to this level is quasibound: classically, it cannot escape, but it can escape via quantum tunnelling and then accelerate in the electric field towards a phosphor screen and CCD camera. The measurement is not adiabatic. The electrons, released from an initial beam of hydrogen atoms via photo-ionization, image the Rydberg wave function on the screen. So, although the experiment involves an ensemble of atoms, each atom contributes independently to the measurement, i.e. reveals a different feature of the initial wave function.

Alongside the experimental development, protective measurements allow us to develop new intuitions for quantum theory. They demonstrate that the members of any ensemble of systems prepared in a given quantum state have much more in common than what previous measurements showed. Not only do all systems prepared in an eigenstate of an operator A yield the same eigenvalue when subjected to a measurement of A; they yield the same expectation values for *any* operator that can be measured on the system. Thus, an ensemble of systems prepared in a given state share a "group identity" which is much richer than a shared eigenvalue: it includes every expectation value that can be measured on the state. The next two sections show that this group identity has implications for the ontology of the quantum state.

10.3 The Pusey–Barrett–Rudolph (PBR) theorem

Probability distributions are often interpreted as subjective, i.e. as representing an observer's knowledge (or ignorance) about a system. Is this also the correct way to interpret probabilities derived from quantum states? There seem to be good reasons to favor such an interpretation, because the alternative interpretation – that the quantum state is no more than a description of the reality of a system – is disturbing in several respects. As a description of reality, the quantum state apparently exhibits instantaneous collapse over unbounded spatial regions. It also superposes properties that are (classically) mutually exclusive. Entanglement implies that the quantum state of a composite system cannot be reduced to states of the component systems. These peculiarities are less troubling if the quantum state represents information about a system, rather than the system's actual physical state [14, 15].

Harrigan and Spekkens [16] gave this question a precise formulation. If the quantum state is a representation of knowledge about an unknown and possibly inaccessible physical state, it does not depend solely on the properties of the system. It depends also on the information available to the observer. Therefore, if the quantum state represents subjective knowledge, at least some *physical* state has to be compatible with more than one *quantum* state. The probabilistic nature of the predictions of quantum theory seems to allow for such compatibility, as long as the two quantum states that can represent one reality are not orthogonal.

More formally, let λ (which could be a number or a vector, and belongs to a space denoted by Λ) be a complete specification of the physical state of a system, e.g. of an atom. If a quantum state $|\Psi\rangle$ of that system corresponds to a single λ, then $|\Psi\rangle$ as well is a complete specification of the physical state of the system. But, in general, $|\Psi\rangle$ could correspond to a probability distribution $p_\Psi(\lambda)$ over the values of λ. If so, then the values of λ play the role of hidden variables of a system in the state $|\Psi\rangle$. Now consider two possible states of the system, $|\Psi\rangle$ and $|\Phi\rangle$. If $|\Psi\rangle$ and $|\Phi\rangle$ are orthogonal, then their respective probability distributions, $p_\Psi(\lambda)$ and $p_\Phi(\lambda)$, must be *non-overlapping*, i.e. $p_\Psi(\lambda)p_\Phi(\lambda) = 0$ for all λ. Otherwise, a prediction of quantum theory – namely, that a measurement of the projection operator $\Pi_\Phi \equiv |\Phi\rangle\langle\Phi|$ on a system prepared in the orthogonal state $|\Psi\rangle$ yields 0 – will fail (since measuring devices respond only to the physical state). But if $|\Psi\rangle$ and $|\Phi\rangle$ are not orthogonal (and not identical), it is conceivable that $|\Psi\rangle$ and $|\Phi\rangle$ could overlap. If $|\Psi\rangle$ and $|\Phi\rangle$ overlap, so that for some $\bar{\lambda}$ we have $p_\Psi(\bar{\lambda})p_\Phi(\bar{\lambda}) \neq 0$, then the two distinct quantum states $|\Psi\rangle$ and $|\Phi\rangle$ could represent the same physical reality $\bar{\lambda}$. Conversely, if $p_\Psi(\lambda)p_\Phi(\lambda) = 0$ for all λ and for any two distinct states $|\Psi\rangle$ and $|\Phi\rangle$, then quantum states represent physical reality.

What is beautiful about this formulation is that it cleanly pulls apart two different questions about the quantum state. The first question – the title of the famous EPR paper [17] and of Bohr's reply [18] – is whether the quantum state is a complete description of a physical state, i.e. whether one quantum state can represent more than one physical state. (If a quantum state $|\Psi\rangle$ completely describes a physical state $\bar{\lambda}$, then $|\Psi\rangle$ cannot represent more than one physical state.) The second question is whether two quantum states can represent one and the same physical state. The Pusey–Barrett–Rudolph (PBR) [19, 20, 21] theorem states that if the predictions of quantum mechanics are correct, then the answer to the second question is negative, regardless of the answer to the first question: no two distinct quantum states can represent the same physical reality, regardless of completeness.

The proof of the PBR theorem is technical. Here we try to motivate the proof intuitively. We begin by assuming that two non-orthogonal but distinct qubit states, $|0\rangle$ and $|+\rangle$, with $\langle 0|+\rangle = 1/\sqrt{2}$, represent in all cases exactly the same physical reality $\bar{\lambda} \in \Lambda$. That is, both $p_0(\lambda)$ and $p_+(\lambda)$ vanish for $\lambda \neq \bar{\lambda}$. Particles in a mixture of the states $|0\rangle$ and $|+\rangle$ are fed into a device that measures an operator with non-degenerate eigenstates $|0\rangle$ and $|1\rangle$, where $\langle 0|1\rangle = 0$. Quantum mechanics predicts that the device should sometimes find a particle in the state $|1\rangle$, but only if the initial state was $|+\rangle$, never if the initial state was $|0\rangle$. But, by assumption, $|0\rangle$ and $|+\rangle$ represent the same physical state $\bar{\lambda}$; hence there is *no way* the device can distinguish them, and, if it finds any particle in the state $|1\rangle$, it must do so sometimes also when the particle's initial state was $|0\rangle$. Thus our assumption implies a violation of quantum predictions.

So far, the proof was easy because we assumed that $|0\rangle$ and $|+\rangle$ can only represent a single physical reality $\bar{\lambda}$. What if $|0\rangle$ and $|+\rangle$ correspond to overlapping distributions $p_0(\lambda)$ and $p_+(\lambda)$? Now the device could find particles in the state $|1\rangle$ *only* for values of λ for which $p_+(\lambda) \neq 0$ and $p_0(\lambda) = 0$, i.e. not in the overlap of the distributions. Hence the device need not violate quantum predictions: it finds the state $|1\rangle$ only when the initial state is not $|0\rangle$. To contradict quantum predictions, the device would have to measure an operator with a non-degenerate eigenstate $|-\rangle$ orthogonal to $|+\rangle$ as well as the non-degenerate eigenstate $|1\rangle$ orthogonal to $|0\rangle$. Of course, no Hermitian operator can have $|1\rangle$ and $|-\rangle$ as non-degenerate eigenstates. What PBR showed, however, is that for a mixture of *pairs* of particles prepared in the states $|0\rangle \otimes |0\rangle$, $|0\rangle \otimes |+\rangle$, $|+\rangle \otimes |0\rangle$ and $|+\rangle \otimes |+\rangle$, there is a non-degenerate operator, on the four-dimensional Hilbert space spanned by these eigenvectors, with the following property: each of these four preparations is orthogonal to one of the operator's eigenstates. Explicitly, the eigenstates are

$$|\xi_1\rangle = \frac{1}{\sqrt{2}} (|0\rangle \otimes |1\rangle + |1\rangle \otimes |0\rangle), \tag{10.3}$$

$$|\xi_2\rangle = \frac{1}{\sqrt{2}} (|0\rangle \otimes |-\rangle + |1\rangle \otimes |+\rangle), \tag{10.4}$$

$$|\xi_3\rangle = \frac{1}{\sqrt{2}} (|+\rangle \otimes |1\rangle + |-\rangle \otimes |0\rangle), \tag{10.5}$$

$$|\xi_4\rangle = \frac{1}{\sqrt{2}} (|+\rangle \otimes |-\rangle + |-\rangle \otimes |+\rangle). \tag{10.6}$$

Now the measuring device cannot avoid violating a prediction of quantum mechanics every now and then. Note that the case of $|0\rangle$ and $|+\rangle$ is special, because without the assumption $\langle 0|+\rangle = 1/\sqrt{2}$ above, the states in Eq. (10.3) would not be orthogonal. The PBR proof of the general case is still more technical.

As an application of the PBR theorem, let us revisit the EPR paper [17]. Consider an entangled state

$$\frac{1}{\sqrt{2}}\left[|0\rangle_A \otimes |1\rangle_B - |1\rangle_A \otimes |0\rangle_B\right] \qquad (10.7)$$

of a particle pair shared by Alice and Bob, far apart in their respective laboratories. If Alice measures $|1\rangle_{AA}\langle 1| - |0\rangle_{AA}\langle 0|$ on her particle, she might leave Bob's particle in the state $|0\rangle_B$; but if she measures $|+\rangle_{AA}\langle +| - |-\rangle_{AA}\langle -|$ on her particle, she might leave Bob's particle in the state $|+\rangle_B$, which is distinct from $|0\rangle_B$ and not orthogonal to it. The state $|0\rangle_B$, claim EPR, must represent the same physical reality as the state $|+\rangle_B$, since no influence, including Alice's measurement, can propagate faster than the speed of light. But according to the PBR theorem, if the predictions of quantum mechanics are correct, then the state $|0\rangle_B$ *cannot* represent the same reality as $|+\rangle_B$. We see that the EPR assumption – according to which Alice's measurement does not disturb Bob's particle – is incompatible with quantum mechanics. It is striking that both Bell's theorem [22, 23] and the PBR theorem imply that EPR's demand for locality (Einstein separability) is incompatible with quantum mechanics, even though the PBR theorem does not mention locality.

10.4 Protective measurement, PBR and the reality of $|\Psi\rangle$

Assuming that quantum predictions are correct, the PBR theorem implies that a quantum state representing an individual system also represents a part or all of the physical reality of that system. Independently, protective measurements make it possible to measure expectation values, including the norm and relative phase of the wave function itself, on an individual system. Since expectation values have physical meaning, the PBR theorem and protective measurements both imply that a quantum state represents physical reality. Would it be right to say that protective measurements anticipate the PBR result? In this section, we show that the answer to this question cannot be a simple Yes or No: although close in spirit, protective measurements and the PBR theorem make different and complementary statements about the physical reality of quantum states. First, however, we address the question of what it means to represent physical reality – a question that is not straightforward in quantum theory.

Hartle [15] claims that the quantum state is not an objective property of the system, because no assertion about the state of the system "can be verified by measurements on the individual system without knowledge of the system's previous history". Indeed, if we are given a single system in an unknown quantum state, protective measurements cannot identify its state, any more than other measurements can. Hartle's conclusion is therefore that a quantum state is a property of an ensemble, but not a property of any individual system. (See also [24].) If so, then neither the PBR theorem not protective measurements make any statement about the reality of the quantum state of a single system.

Hartle's criterion – measurability without prior knowledge – is suitable for the classical world, but it rules out discussion of a single quantum system, and is thus unsuitable for the quantum world. It does not allow attribution of *any* contingent property to individual quantum systems. The quantum world requires a more subtle criterion.

A better criterion for attributing a property to an individual quantum system is that of EPR (italics in the original): "*If, without in any way disturbing a system, we can predict with certainty (i.e. with probability equal to unity) the value of a physical quantity, then there exists an element of physical reality corresponding to this physical quantity.*" We cannot apply this criterion in the case considered by EPR because of the failure of locality, but we can apply it here. For example, if we have just found an isolated atom to be in an eigenstate of energy, we can be sure that a second measurement of its energy will yield the same result, and we can therefore attribute that energy to the atom. Similarly, via protective measurements we can attribute to each quantum system a distinct reality defined by distinct properties. Specifically, protective measurements can be used to probe an individual system in the same way that other quantum measurements probe an ensemble.

How do we show that $|\Psi\rangle$ is a property of an ensemble? We prepare an ensemble of systems in the state $|\Psi\rangle$, and perform measurements on that ensemble. We are not totally ignorant of the preparation; on the contrary, we must know how the state is prepared in order to assign the properties we find to $|\Psi\rangle$. If we were totally ignorant of the preparation, we couldn't even be sure of having an ensemble of identical systems. We can view protective measurements in the same light, but instead of preparing an ensemble of many systems in the state $|\Psi\rangle$, we prepare a single system in the state $|\Psi\rangle$ and protect the state from changing during the (extended) measurement. When we do so, and measure the expectation values of any operator we wish, we obtain values that define the quantum state uniquely. For example, a measurement of a projection operator $|\Psi\rangle\langle\Psi|$ yields 1 in the state $|\Psi\rangle$ and less than 1 in any other state. Since (protective) measurements on any given quantum state yield expectation values that differ from measurements on any distinct quantum state, each quantum state represents a distinct reality.

Thus protective measurements show the reality of a single system in a quantum state. But can they be used to prove that two quantum states inevitably represent different realities? The answer to this question has to be negative, because if two non-orthogonal states $|\Psi_1\rangle$ and $|\Psi_2\rangle$ are possible inputs for a single measuring device, then the possible outputs cannot be orthogonal. Concretely, let us consider a measuring device prepared in a neutral state $|\chi_0\rangle$ and coupled to either $|\Psi_1\rangle$ or $|\Psi_2\rangle$. To discriminate unambiguously between these two states, the measuring device coupled to $|\Psi_i\rangle$ must evolve into the respective pointer state $|\chi_i\rangle$, where $\langle\chi_2|\chi_1\rangle = 0$. But then the initially *non-orthogonal* states $|\chi_0\rangle \otimes |\Psi_1\rangle$ and $|\chi_0\rangle \otimes |\Psi_2\rangle$ must evolve into the *orthogonal* states $|\chi_1\rangle \otimes |\Psi_1\rangle$ and $|\chi_2\rangle \otimes |\Psi_2\rangle$, which is impossible with unitary time evolution.

Thus the description of a quantum state via protective measurements leaves an ambiguity, which we can summarize as follows. Suppose we prepare an ensemble of systems in the state $|\Psi_1\rangle$, protect the state $|\Psi_1\rangle$, and measure a long list of observables. We will, in each case, obtain the expectation values of those observables in the state $|\Psi_1\rangle$. However, if we prepare the ensemble of systems in the state $|\Psi_2\rangle$ (neither identical nor orthogonal to $|\Psi_1\rangle$) and protect the state $|\Psi_1\rangle$, the protection will leave a fraction $|\langle\Psi_2|\Psi_1\rangle|^2$ of the systems in the state $|\Psi_1\rangle$. Now if we measure the same long list of observables, we will obtain a sub-ensemble of systems yielding the expectation values of those observables in the state $|\Psi_1\rangle$. We then cannot eliminate the possibility that the reality λ underlying some of the systems prepared in the state $|\Psi_2\rangle$ is compatible with $|\Psi_1\rangle$ as well as $|\Psi_2\rangle$; we do not have a no-go theorem. One could consider deriving such a theorem by considering, as PBR do, many systems prepared independently in either the state $|\Psi_1\rangle$ or in the state $|\Psi_2\rangle$, but such a derivation would have little to do with protective measurements.

This difference between protective measurements and the conclusion of PBR demonstrates the inherent inaccessibility of quantum reality. We already knew that orthogonal states represent different physical realities. We now know that non-orthogonal states, as well, represent different physical realities. If they represent different physical realities, which observable can we measure to distinguish one physical reality from the other? Of course, there is no such observable; if there were, two non-orthogonal states would be its non-degenerate eigenvectors. Unitarity prevents us from distinguishing two non-orthogonal quantum states, and the PBR theorem implies that this indistinguishability does not arise because two quantum states can represent the same reality. They cannot.

To conclude, both protective measurements and PBR can be seen as partial answers to a single question: what do two quantum systems, described by two quantum states in a common Hilbert space, have in common? Protective measurements tell us that if the two quantum states are the same, the systems have a lot in common, namely the expectation values of all operators measurable on the systems. The PBR theorem tells us that if the two quantum states are different, the systems are in physically different states.

Acknowledgements

We thank Professor Lev Vaidman for Reference [12]. This publication was made possible through the support of grants from the German-Israel Foundation (grant no. 1054/09), from the John Templeton Foundation (Project ID 43297) and from the Israel Science Foundation (grant no. 1190/13). The opinions expressed in this publication are those of the authors and do not necessarily reflect the views of any of these supporting foundations.

References

[1] Y. Aharonov and L. Vaidman, Measurement of the Schrödinger wave of a single particle, *Phys. Lett.* **A178**, 38 (1993).

[2] Y. Aharonov, J. Anandan and L. Vaidman, Meaning of the wave function, *Phys. Rev.* **A47**, 4616 (1993).

[3] J. Anandan, Protective measurement and quantum reality, *Found. Phys. Lett.* **6** (1993).

[4] E. Schrödinger, *Science, Theory and Man* (London: Allen and Unwin), 1957, p. 177.

[5] M. Born, Zur Wellenmechanik der Stossvorgänge, *Gött. Nachr.* (1926), 146, cited and trans. in A. Pais, *Niels Bohr's Times: in Physics, Philosophy and Polity* (New York: Oxford University Press), 1991, p. 286.

[6] Y. Aharonov and D. Rohrlich, *Quantum Paradoxes: Quantum Theory for the Perplexed* (Weinheim: Wiley-VCH), 2005, Chapter 15.

[7] J. S. Lundeen, B. Sutherland, A. Pater, C. Stewart and C. Bamber. Direct measurement of the quantum wavefunction, *Nature* **474**, 188 (2011).

[8] J. S. Lundeen and C. Bamber, Procedure for direct measurement of general quantum states using weak measurement, *Phys. Rev. Lett.* **108**, 070402 (2012).

[9] Y. Aharonov, D. Z. Albert and L. Vaidman, How the result of a measurement of a component of the spin of a spin-$\frac{1}{2}$ particle can turn out to be 100, *Phys. Rev. Lett.* **60**, 1351 (1988); see also Y. Aharonov and D. Rohrlich, [6], Chapters 16–17.

[10] Y. Aharonov, private communication.

[11] A. S. Stodolna, A. Rouzée, F. Lépine et al., Hydrogen atoms under magnification: direct observation of the nodal structure of Stark states, *Phys. Rev. Lett.* **110**, 213001 (2013).

[12] S. Cohen, M. M. Marb, A. Ollagnier et al., Wave function microscopy of quasibound atomic states, *Phys. Rev. Lett.* **110**, 183001 (2013).

[13] D. Lüftner, T. Ulesa, E. M. Reinisch et al., Imaging the wave function of adsorbed molecules, *Proc. Nat. Acad. Sci.* (USA), doi: 10.1073/pnas.1315716110 (2013).

[14] W. Heisenberg, *Physics and Philosophy: the Revolution in Modern Science* (New York: Harper and Row), 1958, p. 54.

[15] J. B. Hartle, Quantum mechanics of individual systems, *Am. J. Phys.* **36**, 704 (1968).

[16] N. Harrigan and R. W. Spekkens, Einstein, incompleteness, and the espistemic view of quantum state, *Found. Phys.* **40**, 125 (2010).

[17] A. Einstein, B. Podolsky and N. Rosen, Can quantum-mechanical description of physical reality be considered complete?, *Phys. Rev.* **47**, 777 (1935), reprinted in *Quantum Theory and Measurement*, eds. J. A. Wheeler and W. Zurek (Princeton: Princeton University Press), 1983, pp. 138–141.

[18] N. Bohr, Can quantum-mechanical description of physical reality be considered complete?, *Phys. Rev.* **48**, 696 (1935), reprinted in J. A. Wheeler and W. Zurek, (see [17]), pp. 145–151.

[19] M. F. Pusey, J. Barrett and T. Rudolph, On the reality of the quantum state, *Nature Phys.* **8**, 475 (2012).

[20] J. Barrett, E. G. Cavalcanti, R. Lal and O. J. E. Maroney, No ψ-epistemic model can fully explain the indistinguishability of quantumm state, arXiv:1310.8302v1 (2013), *Phys. Rev. Lett.* **112**, 250403 (2014).

[21] P. G. Lewis, D. Jennings, J. Barrett and T. Rudolph, Distinct quantum states can be compatible with a single state of reality, *Phys. Rev. Lett.* **109**, 150404 (2012).

[22] J. S. Bell, *Physics* **1**, 195 (1964).

[23] J. F. Clauser, M. A. Horne, A. Shimony and R. A. Holt, Proposed experiment to test local hidden-variable theories, *Phys. Rev. Lett.* **23**, 880 (1969).

[24] L. E. Ballentine, The statistical interpretation of quantum mechanics, *Rev. Mod. Phys.* **42**, 358 (1970).

11

The roads not taken: empty waves, wave function collapse and protective measurement in quantum theory

PETER HOLLAND

Two roads diverged in a wood, and I –
I took the one less traveled by,
And that has made all the difference.
Robert Frost (1916)

11.1 The explanatory role of empty waves in quantum theory

In this chapter we are concerned with two classes of interpretations of quantum mechanics: the epistemological (the historically dominant view) and the ontological. The first views the wave function as just a repository of (statistical) information on a physical system. The other treats the wave function primarily as an element of physical reality, whilst generally retaining the statistical interpretation as a secondary property. There is as yet only theoretical justification for the program of modelling quantum matter in terms of an objective spacetime process; that *some* way of imagining how the quantum world works between measurements is surely better than none. Indeed, a benefit of such an approach can be that "measurements" lose their talismanic aspect and become just typical processes described by the theory.

In the quest to model quantum systems one notes that, whilst the formalism makes reference to "particle" properties such as mass, the linearly evolving wave function $\psi(x)$ does not generally exhibit any feature that could be put into correspondence with a localized particle structure. To turn quantum mechanics into a theory of matter and motion, with real atoms and molecules consisting of particles structured by potentials and forces, it is necessary to bring in independent physical elements not represented in the basic formalism. The notion of an "empty wave" is peculiar to those representatives of this class of extended theories which postulate that the additional physical element is a corpuscle-like entity or point particle. For clarity, we shall develop the discussion in terms of a definite model of this kind

145

whose properties are well understood and which it is established reproduces the empirical content of quantum mechanics: the de Broglie–Bohm theory, a prominent representative of the class of ontological interpretations (Holland, 1993). Here, material physical systems are postulated to consist of two components: a physically real wave (described by $\psi(x)$) governed by Schrödinger's equation, and a point particle that is guided along a track $x(t)$ by the wave (according to the law $m\dot{x} = \nabla S$, where S is the quantal phase) but does not participate in the latter's dynamics (one can extend the model to include a back-reaction of the particle on the wave in a way that is compatible with quantal predictions (Holland, 2006) but this is not needed here). The position of the particle is the "observable" of the theory. Note that this dualistic theory of matter discerns, and attributes ontological significance to, features of the wave function – such as energy and force – that may not be meaningful in other ontological interpretations (which therefore may be incommensurable).

If the ψ-wave is incident on a beam-splitter and evolves into two spatially disjoint components the particle will enter only one of them and the other, by virtue of not containing the particle, will be "empty". It is only in this sense that we shall say a wave is empty – it still propagates energy and momentum of the field throughout space and has the potential to subsequently act on its associated particle if it is finite in a domain where the latter passes (the energy–momentum is only indirectly observable through the effects of the wave on the particle).

To illustrate the active role of an empty wave we recall how it contributes to the de Broglie–Bohm description of the two-slit experiment (Holland, 1993, section 5.1). Referring to Fig. 11.1, a wave $\psi(x)$ incident on a beam-splitter B

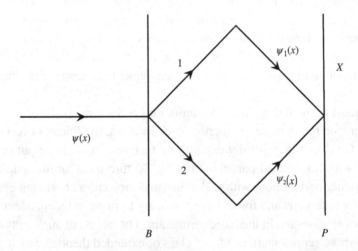

Figure 11.1 A particle traversing path 1 may arrive at an otherwise inaccessible point X due to the action of the empty wave ψ_2.

splits into two packets ψ_1 and ψ_2 which separate sufficiently so that they do not appreciably overlap before being recombined in the vicinity of a screen P. Two distinct routes, 1 and 2, are then available to a particle x passing through the interferometer. If the particle is detected at a point X above the symmetry axis A we know that it traversed route 1 because the single-valuedness of the wave function forbids the crossing of paths. Hence, after the splitting and prior to the recombination, ψ_2 is an empty wave. But the point X may lie in a region not accessible to a particle guided by ψ_1 alone, that is, in the case where ψ_2 is absent (this may be arranged, for example, if ψ_1 possesses nodes where ψ_2 is finite). Hence, the empty wave ψ_2 has had a physical effect in bringing about an observable change in the state of the corpuscle: when ψ_2 is present (absent) the particle can (cannot) land at X.

Notice that the claim that the empty wave has acted physically is a retrospective inference – we argue that it must have so functioned prior to the detection of the particle in order that the result obtained could actually occur. Naturally, the empty wave concept has meaning only within the model of quantum motion we have employed to explain the functioning of the interferometer; the interference phenomenon itself does not prove the "reality" of the empty wave.

It would clearly be advantageous if the historical dispute between the epistemological and ontological viewpoints could be made an empirical issue. In this chapter we examine the impact of the empty wave concept on this problem. We first emphasize the theoretical merits of the empty wave in enabling avoidance of the wave function collapse hypothesis (Section 11.2) and in supplying conceptual precision in the application of quantum mechanics, with particular reference to an example where protective measurements have been used in path detection (Section 11.3). We then go on to address the problem of how the reality of an empty wave might be demonstrated by its effect on other systems, and advance general arguments against this possibility (Section 11.4). However, these arguments are not conclusive and we describe how an alternative perspective in probing the empirical implications of empty waves is provided by the notion of protective measurement (Section 11.5).

11.2 Measurement: empty waves *vs.* wave function collapse

The empty wave concept extends easily to a many-particle system where it is a key characteristic of the configuration space description. An analogue of a beam splitter in configuration space will create a spectrum of waves and the system point will distinguish one of them if they are non-overlapping packets. Note that the physical particles composing the system point need not be located nearby in three-dimensional space.

A drawback of the epistemological interpretation is that it entails the hypothesis that the wave function "collapses" at some stage in a measurement process as the knowledge of the "observer" regarding the state of a system changes, a notion that is hard to formulate unambiguously and consistently (e.g., in relation to relativity). One of the virtues of the de Broglie–Bohm approach is that it provides a coherent account of measurement that, in particular, dispenses with the problematic collapse hypothesis through the use of empty waves.

The measurement problem of quantum mechanics arises when one attempts to attribute definite outcomes to processes devoted to discovering information on a quantum system (Holland, 1993, chapter 8). The measurement of an observable represented by an operator $\hat{A}$ associated with a system having a coordinate x is customarily modelled by an impulsive interaction generated by the Hamiltonian $H = f\hat{A}(-i\hbar\partial/\partial z)$, where z is the coordinate of the apparatus and f is a constant. At first the system and apparatus are non-interacting so the total initial state is $\Psi_0(x, z) = \psi_0(x)\phi_0(z)$, where $\psi_0(x) = \sum_a c_a\psi_a(x)$ is a superposition of eigenstates of $\hat{A}$, and $\phi_0(z)$ is the initial apparatus state (assumed to be a localized packet). The impulsive interaction acts as a beam splitter in configuration space, generating a spectrum of macroscopically distinct apparatus states each correlated with an individual eigenfunction. If the period of interaction is T we obtain

$$\Psi(x, z, T) = \sum_a c_a\psi_a(x, T)\phi_a(z, T), \tag{11.1}$$

where $\phi_a(z, T) = \phi_0(z - faT)$ represents a set of non-overlapping outgoing apparatus packets. These packets are in turn coupled to many-body packets so that their separation is amplified to the macroscopic scale. Each packet corresponds to a possible outcome of the measurement. But the state is a superposition of outcomes and, in order to extract a definite result from the superposition, the hypothesis is invoked in the epistemological interpretation that the state (11.1) "collapses" into one of the summands, say the ath, with probability $|c_a|^2$:

$$\sum_{a'} c_{a'}\psi_{a'}(x, T)\phi_{a'}(z, T) \rightarrow \psi_a(x, T)\phi_a(z, T) \tag{11.2}$$

(after normalization). This transformation is not described by the unitary evolutionary law of quantum mechanics (Schrödinger's equation) and suggestions for how it might come about have ranged from the intervention of an observer who becomes aware of the outcome to modifications of the Schrödinger equation. But, even if it is assumed that it does actually take place, the notion of collapse does not in itself solve the measurement problem. For, to infer the outcome of the measurement, the pointer of the apparatus must be assigned a location whose variation during the interaction can be determined unambiguously. In contrast, according to its usual interpretation, the wave function attributed to the apparatus determines just the statistical frequency of measurement results. The wave function does not itself offer a description of an autonomous moving object. One may attempt to address this difficulty by invoking the feature that $\phi_0(z)$ is sharply peaked about a spacetime orbit, that is, by making some kind of literal identification of the packet with the particle. Then one is tacitly shifting the interpretation of the wave function towards an ontological view, but not in a clearly consistent way – the eventual diffusion of the packet, or the possibility of splitting it into disjoint parts, mean the "particle" does not remain localized, for instance.

Another option is that the projection (11.2) does not take place. Rather, the correct wave function remains (11.1), so that all terms in the superposition continue to be finite, but one is selected as representing the outcome of the measurement because it carries some special attribute. This is the thesis of the de Broglie–Bohm model.

In an ensemble of particle systems the probability density of presence in the initial state is $|\Psi_0(x, z)|^2$. Then, in the measurement, one of the outgoing summands is singled out because the de Broglie–Bohm system point $(x(t), z(t))$ enters it (i.e., it occupies the region where the summand is finite). All the other packets are then empty. In particular, the outcome of the measurement is the position $z(t)$. Since the outgoing packets are non-overlapping, from the standpoint of the *particles* the transformation (11.2) does in effect occur, even though the other ψ_as and ϕ_as are still finite (but empty). The Born probability formula follows since over an ensemble the particle x enters the ath packet with relative frequency $|c_a|^2$. Within this approach, the entire measurement process may be treated by applying the usual linear, unitary Schrödinger equation, and the single concept of particle trajectory enables one to both avoid the collapse postulate and solve the problem of the definiteness of the pointer (and object) position. We shall return to the issue of distinguishing the epistemological and ontological views in this context in Section 11.5.

11.3 The art in quantum mechanics: path detection and conceptual precision

11.3.1 Theory of path detection

The de Broglie–Bohm theory is particularly suited to analyzing the interplay between the observation of interference effects and the determination of the spacetime path of a quantum system. Path determination in such situations often requires establishing that the system lies within a particular spatial region, rather than locating it using a precision position measurement. For example, in an interferometric context the path may lie within one of two distinguishable beams traversing a device. In that and other settings path detection may be achieved by entangling the

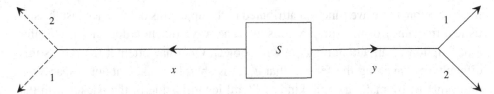

Figure 11.2 A device for determining the path of particle x from the remote detection of particle y using quantum entanglement.

system of interest with another system that has the characteristics of a detector, i.e., that has macroscopically distinguishable states each of which is uniquely coupled with one of the available beams. Here we discuss and attempt to resolve a controversy that has arisen surrounding the application of the ideas of particle trajectory and empty wave in a "which-path" context. It is shown that, if these ideas are applied correctly according to the principles of the de Broglie–Bohm theory, they provide a coherent and uncontroversial account of the functioning of these devices.

An example of such a device arises in the spin $\frac{1}{2}$ version of the EPR experiment. Referring to Fig. 11.2, a source S generates a pair of oppositely moving particles (with magnetic moments) in a singlet state. The particle with spatial coordinate y (the detector) passes through a Stern–Gerlach magnet(s) on the right-hand side oriented in any direction. If y is detected in the upper of the two beams emerging from the magnet (path 1, spin up) then we may infer without further investigation that particle x on the left-hand side pursues the lower beam (path 1), if it subsequently passes through a Stern–Gerlach device oriented in the same direction. Notice that in this example the determination of the path of x has been achieved via a remote local action of the right-hand magnet. Indeed, in this example, the particles never come near each other during the detection process (and the two may be located as far apart as one desires so long as the entangled state is preserved). This is possible because the local action on the magnetic moment of y does not exhaust the dynamical influence of the right-hand magnetic field on the particles, which is mediated also by the wave function. The latter carries information on the local interaction, which is thereby transmitted to y (causing it to move along path 1 or 2) and (nonlocally) to x. It is essential to appreciate that, in this sort of example, the motions of the particles x and y are correlated not because they act upon one another directly as would be expected for two classically interacting particles but because they are each guided by the wave function that carries an imprint of the entire experimental context.

This remote action, whereby a detector locates the de Broglie–Bohm trajectory of a particle with which it does not directly (classically) interact, or even come near, was understood in the context of non-local EPR correlations in the 1980s (see Holland, 1993, section 11.3 and references therein). However, a conceptually

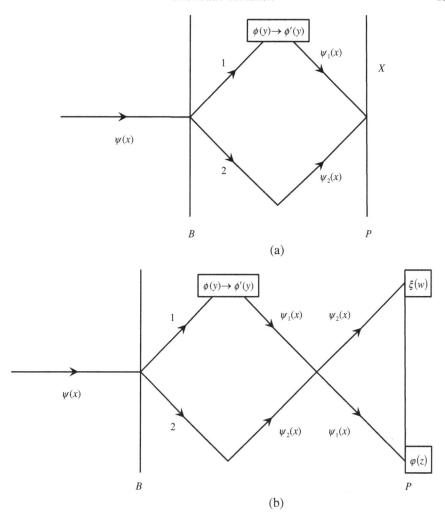

Figure 11.3 The path of particle x inferred from the excitation of detector y depends on the prevailing quantum state: x traverses path 1 in case (a) and path 2 in case (b).

similar example of path detection in interferometry published subsequently has occasioned some (unfounded) disquiet, as we now discuss.

The following is an elaboration of the discussion in Holland (1993, section 8.8). Suppose in Fig. 11.1 we introduce a device in path 1 having wave function $\phi(y)$ and coordinate $y(t)$ (see Fig. 11.3). The purpose of this device is to couple $\psi_1(x)$ and $\phi(y)$ so that the distinct states of the latter allow us to learn along which beam, 1 or 2, x traverses the interferometer. Initially, the total wave function is

$$\Psi(x, z) = [\psi_1(x) + \psi_2(x)]\phi(y) \rightarrow \gamma(x, y) + \psi_2(x)\phi(y) \text{ during interaction in path 1}$$
$$\rightarrow \psi_1(x)\phi'(y) + \psi_2(x)\phi(y) \tag{11.3}$$

after the interaction in path 1, which is assumed to leave $\psi_1(x)$ essentially unaltered. It is required that the initial and final detector states are disjoint in the space of their argument (y), $\phi \cap \phi' = 0$, in order that an unambiguous reading is obtained. Then the two configuration-space summands to which they contribute are disjoint. Thus, if y is found to lie in the excited state $\varphi'(y)$, x must lie in $\psi_1(x)$. The possible outcomes are as follows:

$$y_0 \to y \in \phi'(y) \Rightarrow x \in \psi_1(x),$$
$$y_0 \to y \in \phi(y) \Rightarrow x \in \psi_2(x). \tag{11.4}$$

The outcome in each trial is fixed uniquely by the initial positions x_0, y_0 and the total wave function Ψ. What can we conclude about the path the particle x took through the interferometer?

To bring out how the inference drawn from the meter reading y depends on the total wave function, we consider two possible final wave functions, corresponding to two different experiments that differ in the location of the detecting screen: case (a) when ψ_1 and ψ_2 overlap, and case (b) after ψ_1 and ψ_2 have recombined and, following their natural evolution, passed through one another and separated. Then, from (11.4), when the detector is excited, x may be deduced to pass along path 1 in case (a) and along path 2 in case (b).

It will be noted that, in case (b), x does not pass along the path (1) where the detector is located, that is, an empty wave is associated with the excitation, and the detector locates the particle in a region remote from it. We also find in case (b) that when the detector is unexcited the particle passes through it. Although the details of the devices in Figs. 11.2 and 11.3b differ, they display the same feature of remote detection and for the same reason: correlated motion of the two particles induced by the total wave function that expresses the entanglement of the detector and object and develops into a sum of two disjoint product states in configuration space.

11.3.2 *Realism vs. surrealism*

Englert, Scully and co-workers (Englert *et al.*, 1992) have sought to use these features of scenario (b) to argue that the de Broglie–Bohm theory is not a "realistic" description because the trajectories "may be macroscopically at variance with the observed track of the particle". They present their argument using a Stern–Gerlach interferometer having a detector in each arm but the simpler set-up used in Fig. 11.3b with scalar wave functions and a single detector (suggested by Dewdney *et al.* (1993)) suffices. Their key claim is that for a path detection to occur a detector must fire due to a local interaction between it and the particle whose path is desired, which must be at the detector's location. In an alternative example

(Aharonov *et al.*, 1999), employing a protective measurement (see Section 11.5) to effect a path detection,[1] they suggest that, in the de Broglie–Bohm theory, the momentum transfer to the meter that is involved in the measurement process should be accounted for by the action of the particle x and that, since the particle x in the case they consider does not pass near the relevant point where the interaction "takes place", it cannot have this physical effect. Insofar as their criteria for path detection are not obeyed by the de Broglie–Bohm theory, these authors introduce an artistic metaphor and assert that the trajectories are "surreal". What they seem to mean by this mode of expression is that the trajectories are "wrong". They suggest that the de Broglie–Bohm model needs to specify additional criteria to determine when a legitimate path detection is effected.

Of course, were it the case that a correct description of quantum path detection entailed detectors functioning in the way Englert, Scully and co-workers claim, that is, as involving purely local interactions that reveal the particle trajectory at the location of the detector, this would be an awkward circumstance for the de Broglie–Bohm description. What justifies the claim that a quantum trajectory theory must display such characteristics?

In analyzing this question, the merits of the de Broglie–Bohm approach, in prompting an examination of how language is employed in quantum theory, come to the fore. In the application to protective measurement, Englert, Scully and co-workers justify their claims through appeal to terms like "common quantum sense" and "well localized interaction" as if they are unproblematic components of a clear conceptual framework within which it is legitimate to judge the meaning of the de Broglie–Bohm theory. But the true situation is the inverse of this: the de Broglie–Bohm theory provides the means to assess the worth of a traditional discourse that comprises an extraordinarily vague amalgam of words and concepts tacked on to the quantum formalism. That is, the terms commonly used in quantum mechanics are in fact highly problematic, in particular because they are not part of, or mapped onto, a clear ontology. The purpose of the de Broglie–Bohm theory is precisely to address these shortcomings by providing a consistent framework within which the meanings of terms commonly used in quantal discourse may be assessed. For example, the "conventional view" expounded by these authors that the excitation of a detector functioning through a "well localized interaction" is in itself sufficient to claim that a particle passed through it is unfounded unless supplemented by a physical model consistent with quantum mechanics that allows us to formulate criteria in terms of which it is meaningful to draw such an inference. What is the model of a "particle" for which it can be meaningfully asserted that it "passes"? The conventional view fails to satisfy physicists' natural desire for an

---

[1] Whether a protective measurement can be assimilated to a position measurement has been questioned by Drezet (2006).

unambiguous ontology and for want of an alternative its adherents often slip into a classical discourse for which there is no justification in this context and which, moreover, cannot be consistent. In his paper entitled "Do Bohm trajectories always provide a trustworthy physical picture of particle motion?", Scully (1998) answers in the Abstract: "No. When particle detectors are included particles do not follow the Bohm trajectories as we would expect from a classical type model." And there is the nub of the issue: these critics want interpretations of quantum theory to conform to classical conceptions. A pre-quantum notion of interaction comprising a local exchange of momentum (that has not been proven to be consistent with quantum mechanics) is being invoked to judge a theory (that is proven to be consistent) that indicates how a quite different non-classical notion of "interaction" is necessary if the particle trajectory is to be deployed in a quantum context. In fact, what these authors claim is the "observed track" according to their classical model may indeed not be the actual track based on a quantum model.

As emphasized in relation to the example of Fig. 11.2, in the de Broglie–Bohm model the "interaction" is defined not just by the form of the Hamiltonian but is an action mediated by the configuration space wave function, which implies (non-locally) correlated motions in three-dimensional space. Thus, *local Hamiltonians have non-local effects.* The statement of Aharonov *et al.* (1999) that ". . . an inter-action between the particle and the meter occurs undoubtedly . . ." is the kind of loose language objected to above. When one tries to make this notion precise, as in the de Broglie–Bohm theory, it is seen that it is not a meaningful statement. Rather, one must say that there is an action by the wave function on the two particles which causes them to evolve in a correlated manner so that from the path of one we may infer the path of the other.

Even if particle x travels through the detector when the latter fires, as happens with case (a), there is still no direct interaction between x and y, and *the excitation has not occurred because of the passage.* Indeed, giving significance to the excited as opposed to the unexcited state of the detector is misleading, for in both states one can make an inference as to the path traversed. The fact that the assertion "click = detection of passing particle" is generally unfounded does not so much signal a flaw in the de Broglie–Bohm theory as alert us to the subtlety of the quantum theory of detection that it reveals, in particular that the question of whether or not the particle traverses the detector is *irrelevant* to the issue of path detection. The interaction embodied in (11.3), governed by Schrödinger's equation, occurs whether the wave is empty or not. According to this theory, a path detector *never* directly records the coordinates of the particle "measured" or its "passage".

Indeed, the arbitrary requirement that in a scenario deemed to be one of "path detection" the detected particle must pass in the vicinity of the detector, regardless of the prevailing quantum state, would lead over an ensemble to distributions of readings at variance with those implied by $|\Psi|^2$. In contrast, the de Broglie–Bohm description honours the quantal predictions faithfully. And no additional criteria are needed to specify when a path detection occurs, beyond the reading of the meter y.

In the light of the above we can also assess the analysis of Dewdney *et al.* (1993) who have described the detector as being "fooled" in case (b). In fact, this is somewhat misleading since in all cases the detector performs its function of indicating the particle route; it is no more fooled in this case than in case (a) or in the example of Fig. 11.2.

Distant actions of local interactions are at the heart of the explanatory framework of the de Broglie–Bohm theory and examples abound already in the single particle case. For example, a particle approaching an infinite barrier (the local interaction Hamiltonian) will be reflected without touching it. This happens because the wave function carries information about the local potential (the barrier) to distant points and guides a particle located there. This is not "surreal"; it just shows how quantum theory transcends classical mechanism.

There *is* art in the de Broglie–Bohm picture but it is a subtle, non-classical realism based on a concept of particle interaction for which there is no obvious precedent in pre-quantum physics. The latter aspect is relevant to quantum path detection because of the use of entanglement as a resource. Englert, Scully and co-workers have neglected this feature and hence their criticism is unfounded. One may not care for this aspect of the trajectory theory but to cite it as a blemish in the de Broglie–Bohm description the critic must propose a consistent alternative ontology. As indicated above, in this connection it is not legitimate to invoke as a benchmark "standard quantum mechanics" whose lack of precision was a key motivation for the development of the casual theory in the first place. It has, in fact, often been the lot of the de Broglie–Bohm interpretation to be reproached for seeking to return to classical conceptions (by employing trajectories) only to be faulted for not being classical enough (the trajectories do not do what the critic wants).

11.4 Evidence for empty waves: retrodiction vs. prediction

11.4.1 A general argument against the observability of empty waves

How could the reality of an empty wave be demonstrated? We shall explore here the view that what is desired is a means of manipulating such an entity and its interactions so as to measurably alter the future course of systems that may be potentially influenced by it in a *predictable* way. Two potential methods present themselves. The validity of either method would contradict the hypothesis of wave function collapse. A first method is to bring the empty wave back to influence its own associated particle once we have established the latter's path. Applied to the measurement procedure described in Section 11.2, this would entail reversing the process to achieve overlap of the outgoing macroscopic apparatus packets, a formidable technical challenge. A second method is to try to manipulate the empty wave so as to influence *another* independent wave–particle composite, thus increasing the size of the relevant Hilbert (and configuration) space. An argument has been given (Holland, 1993, section 8.8) that for a general class of interactions the latter method does not allow one to infer the reality of an empty wave, at least according to the criterion of predictability stated above. We now recall this demonstration

Suppose an initial packet $\psi(x)$ containing a particle with coordinate $x(t)$ is split into two packets $\psi_1(x)$ and $\psi_2(x)$ that subsequently separate so that eventually they do not appreciably overlap (Fig. 11.4). The particle will join one or other of the packets. Suppose that ψ_1 interacts with a detector having wave function $\phi(z)$ and coordinate $z(t)$ that can measure the position x, and that subsequently ψ_2 interacts with some other system having wave function $\xi(w)$ and actual location $w(t)$. Initially, the total wave function is

$$\Psi(x, w, z) = [\psi_1(x) + \psi_2(x)] \xi(w) \varphi(z) . \tag{11.5}$$

After the first interaction has commenced the wave function is non-factorizable:

$$\Psi(x, w, z) = \alpha(x, z) \xi(w) + \psi_2(x) \xi(w) \varphi(z) . \tag{11.6}$$

The function $\alpha(x, z)$ is entangled in its configuration space and evolves into a superposition of sharply peaked functions of x (with z-dependent coefficients; see (11.1)). If the corpuscle x lies in ψ_1 it will be found in a region where one of these functions is finite and we then know ψ_2 is an empty wave (since $\psi_1 \cap \psi_2 = 0$).

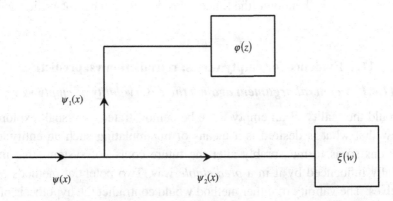

Figure 11.4 An unsuccessful method to detect an empty wave ψ_2 from its effect on a system w.

Now, if we bring in the second interaction, between x and w, the wave function (11.6) becomes

$$\Psi(x, w, z) = \alpha(x, z)\xi(w) + \beta(x, w)\varphi(z) . \tag{11.7}$$

Since the two summands in (11.7) do not overlap, the system point is in one of them. But we have detected that $x(t)$ is in the first summand so we conclude that $w(t)$ cannot be in the second. Hence, because the spatial structure of the function relevant to the motion of the particle w ($\xi(w)$) is unaltered, $\psi_2(x)$ has no observable effect on the particle behavior and we cannot prove the hypothesis of empty waves.

As in the case of Fig. 11.1, what is forbidden is the prediction of an effect of the empty wave once x is detected; we can infer only the past action. As regards the future evolution, empty packets interact only with other *empty* packets – the particles of the other systems are in the same configuration space packet as the particle of interest. All the empty packets do indeed interact with one another and mutually modify their behavior, but this is unobservable since no particles are involved (recall that in this theory the outcomes of experiments are the positions of particles).

This feature of the configuration space dynamics is consistent with experience. If empty waves could really alter the measurable properties of systems, experiments would be constantly perturbed by background noise caused by extraneous ψ-fields and it would be hard to justify the assumption of relative autonomy in which it is legitimate to isolate segments of matter and ignore their environment.

That it is legitimate to draw retrospective inferences, which involve physical elements not included in the predictive apparatus of quantum mechanics, is a feature of some conventional presentations. Thus, Heisenberg (1930) admitted the possibility of reconstructing a trajectory in an interferometer. In both that case and in the examples of empty wave behavior considered here, making correct retrodictions requires adopting a consistent theory of quantum particle motion, or results in contradiction with the statistical distributions of quantum mechanics may be obtained (Holland, 1993, section 8.4.2).

11.4.2 A stronger argument

In the argument just presented against the observability of an empty wave, $\psi_1(x)$ ($\psi_2(x)$) is non-empty (empty) throughout the process. One may envisage more complex scenarios, such as are encountered with beams propagating through an interferometer that separate and then recombine, where these roles may be reversed for periods of the process, prior to the final stage when $\psi_2(x)$ is empty. This introduces the possibility that, in a period when $\psi_1(x)$, say, is empty, we can introduce an interaction in its domain with a detector y so that an outcome of the entire process is that the latter's state alters observably. This raises an issue as to whether this intermediate event could imply evidence for the reality of an empty wave. We shall show that our result remains valid in this more general situation.

In fact, just this circumstance of reversed roles and intermediate interaction with a detector is implicit in the arrangement of Fig. 11.3b and we shall present our proof with reference to that. Including a detector z that interacts with x in the region where $\psi_1(x)$ is finite, the wave function (11.3), in the period after the waves $\psi_1(x)$ and $\psi_2(x)$ have passed through one another and no longer overlap, evolves into

$$\Psi(x, w, y, z) = \alpha(x, z)\, \phi'(y)\, \xi(w) + \psi_2(x)\, \xi(w)\, \phi(y)\, \varphi(z) . \qquad (11.8)$$

If x is detected by the detector z then it lies in the first summand and, since the summands do not overlap, so does w (and y). Therefore, at this final stage the packet $\psi_2(x)$ is an empty wave and an interaction between x and w (which induces $\psi_2(x)\xi(w) \rightarrow \beta(x, w)$) will not influence the future behavior of w, as argued in the case of Fig. 11.4. Hence, we reaffirm our contention that a wave we know to be empty cannot observably influence the future behavior of another physical system.

As anticipated above, the novel element in this example is that over the course of the process the status of $\psi_1(x)$ changes from empty, when it is party (along with the potentials involved in the interaction) to the transformation $\phi(y) \rightarrow \phi'(y)$, to non-empty, when it subsequently interacts with z (in cases where the latter makes a detection). It has been argued by Hardy (1992) in connection with a similar arrangement[2] that, since the then-empty wave ψ_1 is a (partial, in the case he considers) cause of the observable change in y this is evidence that empty waves can "manifest their reality". However, according to our analysis this will not be so. It is true that the empty wave ψ_1 contributed to y's change in state, but because this action occurred *before* the x-detection by z we cannot impute any greater reality to the empty wave here than was possible in the case of, say, Fig. 11.1. Suppose we include in the description of the process depicted in Fig. 11.1 the detection at the screen P, and couple the detector to a light bulb that glows if the detector registers a detection at X. Then we may predict that, when the particle is detected at X, the bulb will glow and this change in its state is caused by the past action of the empty wave ψ_2. The prediction embodied in the change in y in Fig. 11.3b is of a similar type. We only know ψ_1 was empty, in the period when the change occurred, after the z-detection. But by that stage there is no empty wave involved; the relevant empty wave then is ψ_2. It is therefore a retrospective inference that ψ_1 caused

---

[2] The difference between the layouts of Fig. 11.3b and that of Hardy (1992) is the inclusion in the latter of a beam-splitter in the region where ψ_1 and ψ_2 overlap. This introduces an additional interaction so that ψ_1 is only a partial cause of the transformation $\phi \rightarrow \phi'$. Fig. 11.3b is free of this complication.

y's variation. The empty wave certainly contributes to the theoretical account of how the results come about (as analyzed in the de Broglie–Bohm approach) but if the experiment was performed and the quantum predictions confirmed (hardly in doubt) this would not provide evidence for the reality of empty waves.

Although the above results are general and not restricted in the systems to which they apply, there is a caveat: they rest on the assumption that the final interaction between x and w must be localized in a domain of configuration space remote from the region where the first summand is finite. Although this is a natural assumption, it is overly restrictive if we aspire to a comprehensive assessment of all conceivable observable effects of empty waves. Another option is that the system w may interact with both summands in xz-space in a way that maintains their disjointness yet imparts to the first (system-point containing) summand an observable influence depending on the second summand. This possibility is examined next.

11.5 Evidence for empty waves: protective measurement

Progress in attempts to demonstrate an ontological aspect of the wave function came in 1993 when Aharonov and co-workers (Aharonov and Vaidman, 1993; Aharonov *et al.*, 1993; and Anandan, 1993; for reviews and clarifications see Dass and Qureshi, 1999; and Gao, 2013) showed how a suitably adapted adiabatic interaction described by quantum mechanics provides a scheme to measure the expectation values of operators pertaining to a system without appreciably disturbing its quantum state. These interactions are therefore called "protective" measurements. In certain circumstances this provides a technique for "measuring the wave function" of a single system as an extended object (this is not to be confused with the possibility of reconstructing the wave function from a statistical ensemble of conventional measurements (Holland, 1993, section 8.7)). Aharonov *et al.* infer from this procedure, which reveals a property possessed by a single system prior to the measurement, evidence for the ontological character of the wave function. Here we shall point out how the protective measurement protocol, applied in the context of the de Broglie–Bohm model, potentially provides additional support for the ontological viewpoint, by devising a scheme that could distinguish between the empty wave and wave function collapse hypotheses. This application was first suggested by Holland (1994).

We first summarize the theory of protective measurements. Let the initial moment of time be $t = T$ and consider two interacting systems, an object and a measuring apparatus, with initial wave functions $\alpha(x, T)$ and $\beta(y, T)$, respectively. Denote by $\hat{B}$ the operator pertaining to the object whose expectation value is to be measured. Then, in the protective interaction envisaged by Aharonov and co-workers, the interaction Hamiltonian is $H = g(t)y\hat{B}$ and the initial combined state $\Phi(x,y,T) = \alpha(x,T)\beta(y,T)$ evolves adiabatically at time t into

$$\Phi(x,y,t) = \alpha(x,t)\beta(y,t)\exp\left[-(\mathrm{i}/\hbar)\int_T^t g(t)y\langle\hat{B}\rangle\,\mathrm{d}t\right]. \qquad (11.9)$$

Here $g(t)$ is a function characterizing the interaction with $\int_T^t g(t)\mathrm{d}t = 1$, and $\alpha(x,t)$ and $\beta(y,t)$ are the wave functions obtained under free evolution of the two systems. It will be observed that this is still a product state in that the variables x and y have not become entangled. In particular, the object state is undisturbed by the interaction. In contrast, the state of the apparatus has acquired a phase factor, which implies an observable change in its momentum, depending on the expectation value $\langle\hat{B}\rangle = \langle\alpha(t)|\hat{B}|\alpha(t)\rangle$. Hence, information on the state $\alpha(x,t)$ can be gleaned from the apparatus by measuring the change in its momentum. For example, we can choose $\hat{B} = |x_0\rangle\langle x_0|$ so that $\langle\hat{B}\rangle = |\alpha(x_0,t)|^2$ and the shift in momentum is given by $\int_T^t g(t)|\alpha(x_0,t)|^2\mathrm{d}t$, the time-averaged square of the wave amplitude at the point x_0.

If α is known to be a non-degenerate energy eigenstate, but is otherwise unknown, it is possible to use this scheme to measure it for all values of its argument (up to a gauge transformation) by suitable choices of $\hat{B}$. It is to this case that the notion of "measuring the wave function" of a single system using the protective technique really applies. The method may also be applied to general states but there are two caveats: (a) the full Hamiltonian that functions during the protective process depends on the state (Aharonov *et al.*, 1993), which implies that we must first know α before we can investigate it, and (b) that investigation reveals results about time averages of functions of the wave function rather than instantaneous values. So, in the general case the protective scheme provides a way to confirm our time-averaged prior information empirically. Our application of the protective technique below falls into this category; when the protective interaction commences the wave function is the result of a known state preparation procedure (a conventional measurement process). The aspect of the protective process that is important here is that any finite portion of the wave function of interest (obtained by varying its argument) has a discernible effect on a measuring device.

By extending the range of the label x the scheme may be generalized in a straightforward way to provide a protective measurement of a many-particle system, which again may be applied in principle to any wave function. The formula (11.9) remains the same if a single-component observable $\hat{B}$ is measured. It has been argued that this extension supports the attribution of ontological significance to the wave function in configuration space (Anandan, 1993).

We propose to apply this method to the wave function (11.1) that results from a typical conventional measurement in the case where it is assumed the collapse (11.2) does not occur, that is, when (11.1) comprises the set of empty waves generated by the measurement interaction in addition to the one corresponding to the actual outcome. The initial wave function is then the function (11.1); this is the wave function to which the protective interaction is applied (so that we replace x above by x and z). Let us suppose that the configuration point (x, z) of the de Broglie–Bohm model lies in the ath summand of $\Psi(x, z, T)$ and that we determine this fact, and hence the location of particle x, by registering z. Then the other summands are finite but empty from the moment the summands separate and remain so independently of the registration of x. Hence, we may attempt to apply the technique of Aharonov *et al.* to measure functions of the finite, empty components of the total wave function and so provide empirical support for their reality. To this end, fix attention on the a'th component, $a' \neq a$, choose a point $(x_0, z_0) \in \psi_{a'}(x, t)\phi_{a'}(z, t) \neq 0$ (so that $\psi_{a''}(x_0, t)\phi_{a''}(z_0, t) = 0$ for all $a'' \neq a'$), and let the operator pertaining to the object (here the first object plus first detector) be $\hat{B} = |z_0\rangle|x_0\rangle\langle x_0|\langle z_0|$. Then, from (11.9) we obtain

$$
\Psi(x, y, z, t) = \left[\sum_a c_a \psi_a(x, t)\phi_a(z, t) \right] \beta(y, t)
$$

$$
\times \exp\left[-(i/\hbar)y \int_T^t g(t) \left| \psi_{a'}(x_0, t)\phi_{a'}(z_0, t) \right|^2 dt \right].
$$

(11.10)

To test whether in the first (conventional) measurement the wave function has really collapsed then requires observing the momentum of the detector y, which in the ontological interpretation has shifted by an amount depending on the finite amplitude of the empty wave $\psi_{a'}(x, t)\varphi_{a'}(z, t)$.

This technique provides an alternative to the first method of testing for empty waves mentioned at the start of Section 11.4 in that it is not necessary to get an empty wave to overlap with the packet containing the system point. It is also not necessary to seek to observe the superposition of outgoing states since we need only select one for attention. There are, however, formidable difficulties of implementation. Some of the practicalities of the protective scheme, such as contamination by entanglement and the problem of measuring the variation in the meter's state, have been discussed in the above references, particularly by Dass and Qureshi (1999). A significant issue for the above proposal is that, for the selected outgoing configuration space packet, the protective scheme is being applied to a *macroscopic* object (through the detector coordinate z and its coupling to further many-body systems). A possible arena in which to apply the scheme is that of the cases studied in Section 11.4. During a protective measurement, the wave function (11.6) evolves into

$$
\Psi(x, w, z) = \alpha(x, z)\xi(w) e^{iwp(\psi_2, \varphi)/\hbar} + \psi_2(x)\varphi(z)\xi(w) e^{iwp(\psi_2, \varphi)/\hbar} \quad (11.11)
$$

rather than (11.7). The system point remains in the first summand in (11.11) and the detector coordinate w acquires a momentum p depending on the empty wave amplitude $\psi_2(x)\varphi(z)$. A similar result is obtained with the wave function (11.8)

It should be pointed out that a less stringent interaction than that of the protective scheme may suffice for the purpose of observing the effect of an empty wave. The key attribute used here is the ability to probe the quantum state as an extended object; it may be permissible to allow the probed state to be modified by some "quasi"-protective interaction, for example.

11.6 Conclusion

We have considered an aspect of the problem of how evidence may be gained to support the contention that a quantum system has a particle component. The empty wave, a concomitant of the particle model, is a useful theoretical element but in generic situations its influence can at best be inferred retrospectively. This is consistent with the fact that empty waves do not generally disturb physical systems. But there are exceptions and we have described how this issue could potentially be brought into the experimental arena when the special conditions of the procedure used in a protective measurement are satisfied. Distinguishing collapse from non-collapse models is theoretically feasible but technically demanding and the challenge is to find a practical implementation of the protective method.

References

Aharonov, Y. and Vaidman, L. (1993). Measurement of the Schrödinger wave of a single particle. *Phys. Lett. A*, **178**, 38.

Aharonov, Y., Anandan, J. and Vaidman, L. (1993). Meaning of the wave function. *Phys. Rev. A*, **47**, 4616.

Aharonov, Y., Englert, B.-G. and Scully, M. O. (1999). Protective measurements and Bohm trajectories. *Phys. Lett. A*, **263**, 137; Erratum. *Phys. Lett. A*, **266**, 216 (2000).

Anandan, J. (1993). Protective measurement and quantum reality. *Found. Phys. Lett.*, **6**, 503.

Dass, N. D. H. and Qureshi, T. (1999). Critique of protective measurements. *Phys. Rev. A*. **59**, 2590.

Dewdney, C., Hardy, L. and Squires, E. J. (1993). How late measurements of quantum trajectories can fool a detector. *Phys. Lett. A*, **184**, 6.

Drezet, A. (2006). Comment on "Protective measurements and Bohm trajectories" [*Phys. Lett. A* **263** (1999) 137]. *Phys. Lett. A*, **350**, 416.

Englert, B.-G., Scully, M. O., Süssmann, G. and Walther, H. (1992). Surrealistic Bohm trajectories. *Z. Naturforsch.*, **47a**, 1175.

Frost, R. (1916). *The Road Not Taken and Other Poems*. New York: Dover (1993 reissue of *Mountain Interval*. New York: Henry Holt & Co. 1916).

Gao, S. (2013). Protective measurement: a paradigm shift in understanding quantum mechanics. Preprint.

Hardy, L. (1992). On the existence of empty waves in quantum theory. *Phys. Lett. A*, **167**, 11.

Heisenberg, W. (1930). *The Physical Principles of the Quantum Theory*. Chicago: University of Chicago Press (reprinted by Dover, New York, 1949).

Holland, P. R. (1993). *The Quantum Theory of Motion*. Cambridge: Cambridge University Press.

Holland, P. R. (1994). Testing wavefunction collapse. Preprint.

Holland, P. R. (2006). Quantum back-reaction and the particle law of motion. *J. Phys. A: Math. Gen.*, **39**, 559.

Scully, M. O. (1998). Do Bohm trajectories always provide a trustworthy physical picture of particle motion? *Phys. Scr.*, **T76**, 41.

12

Implications of protective measurement on de Broglie–Bohm trajectories

AURÉLIEN DREZET

12.1 Motivation

Protective measurements, which were defined by Aharonov and Vaidman in 1993 [1], played an important role in the discussion about the interpretation of quantum mechanics. In 1999, following an early work by Englert et al. [28], Aharonov et al. [5] wrote an article in which they showed that protective measurements can be used to demonstrate the "surrealism" of Bohmian mechanics. Bohmian mechanics, also known as the pilot-wave interpretation, is certainly the best-known hidden variable interpretation of quantum mechanics. It played a fundamental role in the discovery by Bell of his famous non-locality theorem. Therefore, any attacks against pilot-wave interpretation are particularly interesting and instructive to teach us something new about the mysterious quantum universe. It is the aim of this chapter to review the debate surrounding protective measurement and pilot-wave theory (see also [22] and [29]) and to see if it is possible to reconcile the different interpretations of the results given in [5].

12.2 A historical review of the pilot-wave interpretation

We first remind the reader of some basics about the de Broglie–Bohm "pilot-wave" ontology and in particular about its curious history. De Broglie proposed his approach to quantum mechanics in the period 1925–1927, i.e., at the beginning of modern quantum physics as we know it. De Broglie based his interpretation mainly on relativistic considerations and discovered along this path what is nowadays known as the "Klein–Gordon" equation:

$$\Box\Psi(\mathbf{x}, t) = -\frac{m_0^2}{\hbar^2}\Psi(\mathbf{x}, t) \tag{12.1}$$

What is, however, puzzling is that the first calculations he did on this subject in 1925 [16] were realized before the discovery by Schrödinger of his famous equation. In some way, we can therefore say that it is quantum wave mechanics which was a development of pilot-wave theory and not the opposite [7].

More precisely, the starting idea of de Broglie [17] was that each single quantum object is actually some highly localized singularity of a specific wave field $\Psi(\mathbf{x}, t)$ which should ultimately be solution of a yet unknown non-linear wave equation. Following Einstein, who had already proposed similar ideas in 1909 [24] for photons (the so-called "Nadelstrahlung" concept), de Broglie started a research program called "double solution" [19] in which each quantum is some "bunched" oscillating region of the field propagating as a whole like a particle (i.e. in modern words: a soliton) and inducing a much weaker wave field in its surroundings. This weaker field was supposed to be in "harmony of phases" with the singular field so that both were locked to each other. Following this program the weaker field should obey, far away from the core, a linear equation, e.g., Eq. (12.1), and subsequently should act as a guiding or pilot wave for the singular part, i.e., determining its complete dynamics. This was of course a very ambitious project and not surprisingly de Broglie never succeeded in completing his theory [20]. Still, during his early quest in 1927 he found a "minimalist solution" which is the foundation of what we call nowadays the de Broglie–Bohm interpretation of quantum mechanics. The theory was introduced at the end of a long article about his double solution program [19] and was subsequently presented during the fifth Solvay congress which took place in Brussels (see [42], pp. 105–132). In pilot-wave mechanics, the wave is everywhere reduced to its linear contribution, e.g., a solution of the Schrödinger equation in the non-relativistic regime. The particle behaves like a point-like object whose motion is completely determined by the linear wave. De Broglie was able to define the equation of motion of the moving point-like particles (for the single and many electron cases) and showed how to solve the dynamic equation for some specific problems.

Consider for example a single electron described by Schrödinger's equation:

$$i\hbar \frac{\partial}{\partial t} \Psi(\mathbf{x}, t) = \frac{-\hbar^2}{2m_0} \Delta \Psi(\mathbf{x}, t) + V(\mathbf{x}, t). \tag{12.2}$$

If we know a solution of this equation written in polar form as $\Psi(\mathbf{x}, t) = a(\mathbf{x}, t)e^{iS(\mathbf{x}, t)/\hbar}$ we can define a density of probability $\rho(\mathbf{x}, t) = \Psi(\mathbf{x}, t)\Psi(\mathbf{x}, t)^*$, i.e., $\rho(\mathbf{x}, t) = a(\mathbf{x}, t)^2$, and a probability current $\mathbf{J}(\mathbf{x}, t)$ such as

$$\mathbf{J}(\mathbf{x}, t) = \hbar \frac{\Psi(\mathbf{x}, t)^* \nabla \Psi(\mathbf{x}, t) - \Psi(\mathbf{x}, t) \nabla \Psi(\mathbf{x}, t)^*}{2im_0} = a(\mathbf{x}, t)^2 \frac{\nabla S(\mathbf{x}, t)}{m_0}. \tag{12.3}$$

Using these equations de Broglie defined the velocity of the particle as

$$\mathbf{v}(t) = \frac{\mathrm{d}}{\mathrm{d}t} \mathbf{x}(t) = \frac{\mathbf{J}(\mathbf{x}, t)}{\rho(\mathbf{x}, t)} = \frac{\nabla S(\mathbf{x}, t)}{m_0}, \tag{12.4}$$

showing that in analogy with classical dynamics $S(\mathbf{x}(t), t)$ plays the role of an action (see also Madelung [35]). This analogy is even enforced when we insert a and S in Eq. (12.2) to obtain

$$-\frac{\partial}{\partial t}S(\mathbf{x}(t), t) = \frac{(\nabla S(\mathbf{x}, t))^2}{2m_0} + V(\mathbf{x}(t), t) - \frac{\hbar^2 \Delta a(\mathbf{x}(t), t)}{2m_0 a(\mathbf{x}(t), t)}. \tag{12.5}$$

We recognize the well-known Hamilton–Jacobi equation which in classical dynamics determines the motion of the particle in an external potential V. However, there is here an additional term $Q(\mathbf{x}, t) = -\hbar^2 \Delta a(\mathbf{x}, t)/(2m_0 a(\mathbf{x}, t))$ called the quantum potential by de Broglie. This potential is determined by the wave amplitude in agreement with the pilot-wave idea. Importantly, Q is unchanged if the wave function is multiplied by a constant so that actually it is the form of the wave more than its amplitude which has significance in this theory. Also, for a many-body system the potential $Q(x_1, x_2, \ldots, x_N, t)$ depends in general in a non-local way on the N particle coordinates. This can lead to some specific features such as non-local entanglement, discussed in the context of the EPR paradox [26] or the Bell inequality [9]. In particular, the fact that pilot-wave theory agrees with Bell's theorem implies some kind of mysterious action at a distance between the particles. We point out that de Broglie, contrary to Bohm, was very reluctant to introduce non-locality in his ontological theory and that he expected to remove this feature with his double solution program.

Importantly, the Hamilton–Jacobi analogy suggests that pilot-wave theory can equivalently be written in Newton's form. The second law for de Broglie's dynamics is indeed easily written as $m_0 \frac{d^2}{dt^2}\mathbf{x}(t) = -\nabla[V(\mathbf{x}(t), t) + Q(\mathbf{x}(t), t)]$ in full analogy with classical dynamics for a point-like particle. However, while this dynamical law contains a second-order time derivative it is important to observe that for practical purposes if $\Psi(\mathbf{x}, t)$ is known then the first-order Eq. (12.4) is sufficient to completely describe the trajectories. This is indeed done through integration of the flow equations:

$$\frac{dx}{\frac{\partial S(\mathbf{x}, t)}{\partial x}} = \frac{dy}{\frac{\partial S(\mathbf{x}, t)}{\partial y}} = \frac{dz}{\frac{\partial S(\mathbf{x}, t)}{\partial z}} = \frac{dt}{m_0} \tag{12.6}$$

for a given initial condition $\mathbf{x}(t_0) = \mathbf{x}_0$. This point is important because John Bell [9] used pilot-wave theory mainly through the definition given by Eq. (12.4) while other authors like Bohm and Vigier [15] and Bohm [11] insisted on the need to use the quantum potential for a complete physical description of the particle's motion. This seems to indicate that the theory lacks a univocal axiomatic for this foundation.

At the Solvay conference W. Pauli was probably the most reactive concerning criticisms, but even potential followers like Einstein or the more "classical"

Lorentz were not showing a too strong enthusiasm for the de Broglie pilot-wave approach. Remarkably, due to internal mathematical difficulties of his "double solution" program de Broglie only presented his pilot-wave version in Brussels. This was certainly an honest choice but physically far less profound and less impressive for this demanding audience. In particular, one of Pauli's objections concerned the arbitrariness of the dynamics law obtained by de Broglie. Indeed, Pauli observed (see [42], pp. 134–135) that the dynamics proposed by de Broglie has no precise foundation since the conservation current is not univocal, i.e. one can add a divergence-free vector to $\mathbf{J}$ without changing the conservation law, and Schrödinger asked further why we should not use instead of Eq. (12.4) a different definition (see [42], p. 135), e.g., the energy–momentum tensor $T^{\mu\nu}(\mathbf{x}, t)$ in order to define a trajectory. This was indeed proposed by de Broglie himself for photons [18] (see, however, [34] for a modern perspective concerning this problem and the difficulties about a covariant generalization of pilot-wave theory).

We also mention related critical comments concerning this foundation made in 1952 by Pauli [38] and in 1955 by Heisenberg [33]. Both physicists indeed complained by observing that for de Broglie and Bohm the particle position plays a fundamental role that breaks the accepted symmetry between position and momentum (symmetry which is at the heart of quantum formalism). This was unacceptable for Heisenberg and Pauli, for whom position q and momentum p should be introduced an equal footing.

Of course, all these observations by Pauli, Schrödinger and Heisenberg are not decisive remarks against the pilot-wave interpretation since the plausibility or implausibility of the dynamics doesn't constitute by itself a proof or disproof of the theory: only experiments should have the last word. Nevertheless, together these problems gave a strong feeling of discomfort to the audience of the Solvay conference and to the first generations of quantum theorists. This discomfort never really disappeared until recently.

12.3 The measurement theory and the adiabatic theorem

12.3.1 Einstein's reaction

Beyond these interesting problems about axiomatics and foundations the most critical part of the theory concerns of course its agreement with experimental facts and the realism of the predictions given by the pilot-wave approach. Indeed, if Schrödinger's equation completely determines the particle motion through Eq. (12.4) then we expect that both the usual "Copenhagen" approach and the one of de Broglie should be experimentally equivalent. This was indeed later confirmed after the more detailed studies of measurement processes by David Bohm in 1952 [12]. Still, in 1927 de Broglie [19] already showed that the four-vector current J^μ, which naturally arises from wave equation (12.1) and formally leads to the conservation law $\partial_\mu J^\mu = 0$ through the Noether theorem, can be used to justify the statistical interpretation of quantum mechanics, i.e., the so-called "Born's probability rule". Indeed, if for simplicity we limit ourselves to the non-relativistic regime, the evolution equation (12.4) and the current conservation rule $\partial_t \rho + \nabla \cdot \mathbf{J} = 0$ imply the following: if at a given time the probability distribution of a particle in space is given by a^2 then this will also be true at any time. If we write $a^2(\mathbf{x}(t_0), t_0)$ for the density of probability at $\mathbf{x}_0 = \mathbf{x}(t_0)$ and time t_0 we can obtain by direct integration the density of probability at time t for the point $\mathbf{x}(t)$ located along the de Broglie trajectory (see Eqs. (12.4) and (12.6)). We get

$$a^2(\mathbf{x}(t), t) = a^2(\mathbf{x}(t_0), t_0) \cdot e^{-\int_{t_0}^t dt' \frac{\Delta' S(\mathbf{x}(t'), t')}{m_0}}, \tag{12.7}$$

where $\Delta' = \partial^2/\partial \mathbf{x}(t')^2$. This is the same reasoning which is used in classical statistical mechanics, e.g., Liouville and Gibbs, to justify probability laws. In particular, if at a given time the wave functions of particles can be approximated by uncorrelated plane waves then an "a priori" symmetry implies the homogeneity of the probability distribution in space (this can be seen as an initial chaotic condition à la Boltzmann, i.e., a "Stosszahlansatz"). The subsequent interaction processes between the different particles will certainly create correlations between them but then the deterministic evolution (e.g., Eq. (12.4) and its generalization for the many-body problem) will maintain the probability interpretation for any other time t as we already said. This idea was further developed by Bohm [13], Bohm and Vigier [15] and Nelson [37] in the 1950–1960s and more recently by Valentini [44, 45] and Dürr, Goldstein and Zanghì [23] with different strategies.

This is certainly impressive, or at least promising, but the theory possesses some other "repellant" features which were studied in recent years and are the subject of the present chapter. One of them (already mentioned by Ehrenfest in [42], p. 136) concerns the fact that in the ground state of a hydrogen atom (i.e. an s state) the wave function is (up to the $e^{-iEt/\hbar}$ contribution) real. It implies that $\mathbf{v} = \nabla S/m_0 = 0$ i.e, the fact that the electron is at rest in the s-atom. From the point of view of the de Broglie theory there is nevertheless no contradiction since the constant energy E is given by $E = -\partial_t S = V(\mathbf{x}) + Q(\mathbf{x})$ and the variation of Q with $\mathbf{x}$ exactly compensates the variation of V. The force $\mathbf{F} = -\nabla[V + Q]$ therefore vanishes and the electron is not accelerated. Still, this feature looked not realistic and played again against de Broglie. Not surprisingly, after this period 1927–1928 de Broglie abandoned his theory and went back to it only after 1952 and the rediscovery by Bohm of pilot waves. We point out that the "$\mathbf{v} = \nabla S/m_0 = 0$" objection played also a role in the "cold" reception of this theory by Pauli [38] and Einstein even after

1952 (see also Rosen [39] who re-discovered, after de Broglie but before Bohm, the pilot-wave concept and repudiated it for the same reasons as Einstein). In a paper written for Max Born's retirement from the University of Edinburgh [25] Einstein discussed the example of a particle in an infinite 1D potential well which admits wave functions

$$\Psi(x,t) = \sqrt{\frac{2}{L}} \sin(n\pi x/L) e^{-iE_n t/\hbar} \qquad (12.8)$$

associated with the energy $E_n = (\hbar n\pi/L)^2/(2m_0)$ for $n = 0, 1, 2$, etc. Clearly, here again the velocity of the particle cancels. For Einstein this seemed to contradict the fact that for large n an ontological theory like pilot-wave should "intuitively" recover classical mechanics. However, in classical mechanics we have $Q = 0$ and $E = p^2/(2m_0)$ with $p = m_0 v$. This apparently fits well with the Schrödinger equation if we write $p = \hbar n\pi/L$. Unfortunately, the pilot-wave theory of de Broglie and Bohm implies $p = m_0 v = \nabla S = 0$ and $Q = (\hbar n\pi/L)^2/(2m_0) = E$. Most remarkably, this occurs independently of how large the quantum number n is and is therefore in complete contradiction with what we intuitively expect in the classical regime. Commenting further on Bohm's attempt to reintroduce pilot-wave theory Einstein once wrote to Born: "That way seems too cheap to me". Still, we point out that neither de Broglie nor Bohm agreed with Einstein's conclusion. For example, in his book written with B. Hiley [14] Bohm replied that, independently of the details of pilot-wave theory, any model attempting to preserve the particle localization in the infinite potential well would ultimately contradict classical physics. This should be the case since at fixed energy there are necessarily some nodes where the wave function cancels and are therefore prohibited in the particle localization, i.e., corresponding to regions where the probability is zero. In the 1D case the potential well is thus obviously separated into small spatial cells of size $\lambda/2 = L/n$ where the particle is confined and cannot escape because it cannot cross or even reach the nodes. Therefore, in this context the expectation of Einstein appears illusory. Still, the example of Einstein or the one of the s-atom constitute perfect illustrations of the "surrealistic nature of the de Broglie–Bohm trajectory". This qualifier was given by Englert *et al.* after a very detailed paper [28] which discussed the pilot-wave interpretation in the context of measurement theory.

12.3.2 *Von Neumann's strong measurements*

The most important contribution of David Bohm to pilot-wave theory concerns his interpretation of quantum measurements. In 1952 in a series of two well-known papers [11, 12], he discussed the canonical von Neumann projective measurements in the context of pilot-wave theory. He showed that there is nothing contradictory or impossible in attributing at the same time a position and a momentum to a particle as soon as we accept that the so-called momentum measured is not in general its actual momentum. This should be already clear from the definition $\mathbf{p}(t) = m_0\mathbf{v}(t) = \nabla S$ which holds at any location $\mathbf{x}$ visited by the particle. The plane-wave eigenstates $|\mathbf{p}\rangle$ of the operator $\hat{\mathbf{p}} = -i\hbar\nabla$ are completely delocalized and according to the Heisenberg principle this prohibits a clean localization of the particle. The agreement between the definition of de Broglie–Bohm on the one side and of Heisenberg on the other side is re-established if we realize that in order to measure the momentum associated with the operator $\hat{\mathbf{p}}$ one must disturb the initial wave function and separate the different plane wave contributions.

Consider once again the example of the infinite potential well. Bohm observes that the wave function given by Eq. (12.8) can be formally expanded into plane waves and the Fourier amplitudes $\tilde{\psi}(p)$ correspond to two well-localized wave packets peaked near the momentum values $p = \pm\hbar n/L$ (in the classical limit $n \to +\infty$ we have $|\tilde{\psi}(p)|^2 \simeq [\delta(p - \hbar n/L) + \delta(p + \hbar n/L)]/2$). Bohm then supposed that the walls or the well confining the particle instantaneously disappear without disturbing the wave function in any appreciable fashion. The two wave packets subsequently propagate freely and ultimately separate from each other. This makes the packets spatially distinguishable and allows for a measurement of the particle momentum $p \simeq \pm\hbar n/L$.

This example is of course a "gedanken" experiment and subsequent studies made by Bohm and followers focused on the procedure of entanglement between a pointer or meter and the analyzed quantum system.

This was first done in the context of von Neumann measurement, whose method was well discussed by Bohm himself in his "orthodox" 1951 text book, e.g., for the Stern–Gerlach experiment analysis [10]. The main idea can be easily illustrated by considering the total Hamiltonian

$$\hat{H}(t) = \hat{H}_S + \hat{H}_M - \hbar g(t)\epsilon\hat{A}_S\hat{X}_M, \qquad (12.9)$$

describing the interaction between a system S and a meter M. The operator $\hat{A}_S$ acts only on S and corresponds to the variable we wish to measure. $\hat{X}_M$ is the operator describing the meter. It represents here its position (e.g. the atom center of mass in the Stern–Gerlach experiment). The coupling is also characterized by a constant ϵ introduced for the sake of the equation's homogeneity and a time-dependent function $g(t)$ characterizing the fast evolution of the measurement protocol. Here, we impose for simplicity $g(t) = \delta(t)$, i.e., an instantaneous measurement. Before the interaction occurs at $t = 0$ we start, i.e., for $t < 0$, with two decoupled and unentangled subsystems S and M described by the quantum state $|\Psi_{in}(t)\rangle = |S(t)\rangle \otimes |M(t)\rangle$. After the interaction occurs, i.e. for $t > 0$, we obtain the final state $|\Psi_f(t)\rangle = \hat{U}_S(t, t = 0)\hat{U}_M(t, t = 0)|\Psi_f(0)\rangle$, where $\hat{U}_S(t, t = 0)$ and

$\hat{U}_{\mathrm{M}}(t, t = 0)$ are the evolution operators of the freely moving subsystems S and M acting on

$$|\Psi_{\mathrm{f}}(0)\rangle = \mathrm{e}^{\mathrm{i}\epsilon\hat{A}_{\mathrm{S}}(0)\hat{X}_{\mathrm{M}}(0)}|S(0)\rangle \otimes |M(0)\rangle$$

$$= \sum_a \int \mathrm{d}p S(a) M(p) \mathrm{e}^{\mathrm{i}\epsilon a\hat{X}_{\mathrm{M}}}|a\rangle \otimes |p\rangle \tag{12.10}$$

or equivalently on

$$|\Psi_{\mathrm{f}}(0)\rangle = \sum_a \int \mathrm{d}p S(a) M(p)|a\rangle \otimes |p + \hbar\epsilon a\rangle$$

$$= \sum_a \int \mathrm{d}p S(a) M(p - \hbar\epsilon a)|a\rangle \otimes |p\rangle. \tag{12.11}$$

Here we used the expansion of the initial wave packets in the vector basis $|a\rangle$ and $|p\rangle$ respectively and applied well-known properties of the translation operators $\hat{T}(\hbar a) = \mathrm{e}^{\mathrm{i}\epsilon a\hat{X}_M}$. In particular, if we take $M(p) = \mathrm{e}^{-\Delta p^2}$ we obtain after the measurement a series of shifted Gaussians $M'(p) = \mathrm{e}^{-\Delta(p - \hbar\epsilon a)^2}$ entangled with each state $|a\rangle$. If the shift of each Gaussian is larger than their typical width $\delta p = 1/(2\Delta)$ (and if we can neglect the free space spreading of the pointer wave packets) it will be possible to correlate the distribution $|S(a)|^2$ of S with the distribution of Gaussian centers in the momentum space of M. This is the basis of the von Neumann measurement protocol which was translated into the ontological language of pilot-wave theory by Bohm in 1952. For Bohm indeed, the entanglement directly affected the particle trajectories of the two subsystems S and M but if we observe the meter at a location near $p \simeq \hbar\epsilon a$ it does not, however imply that S is actually in the state a. This apparently paradoxical result comes from the fact that in pilot-wave theory the position of particles plays a more fundamental role than in the usual interpretation. Therefore, we should be authorized to speak about measurement only if we can correlate the studied variables a with the actual position of the system S. Interestingly, both interpretations by von Neumann and Bohm of the previous protocol will, however, eventually agree if the different wave packets of the subsystem S: $\psi_a(x_{\mathrm{S}})$ in the base a, are not spatially overlapping. In a more general way, if the entanglement between the system S and meter M produces after the interaction a sum of entangled states $\sum_i c_i\psi_i(x_{\mathrm{S}})\phi_i(x_{\mathrm{M}})$, where the different wave functions for both particles are non-overlapping, we will then unambiguously be able to correlate the positions of S and M with the states labeled by i. For most experiments this is, however, not the case and the so-called quantum measurement cannot be considered as such in the context of pilot-wave theory. It is therefore amazing to observe that the famous dictum of Wheeler, "No elementary phenomenon is a phenomenon until it is an observed phenomenon", which was given in the context of Bohr's interpretation, finds also its plain significance in the interpretation of de Broglie and Bohm. Paraphrasing Wheeler, we could then state that "No measurement is a measurement until it is a position measurement". It is also worth mentioning that in the same texts quoted previously, both Pauli [38] and Heisenberg [33] criticized this strange feature of the pilot-wave approach. Heisenberg, in particular, pertinently commented that in the deterministic approach of Bohm momentum and position are in general hidden and correspond therefore to metaphysical superstructures without any physical implication.

12.3.3 Protective measurements

The previous discussion done in the context of orthodox von Neumann strong projective measurements was extended in 1999 to the so-called weak protective measurement domain by Aharonov, Englert and Scully in a fascinating paper [5]. The authors showed that in the considered regime the interpretation by pilot-wave theory of the results implied some even more drastic surrealism than in the strong coupling regime. To understand their motivation it is important to go back once again to the origin of pilot-wave mechanics and to observe that if the wave guides the particle during its motion then in some situations empty waves without particles should exist. For example, in the double-slit experiment the particle travels through one hole but something should go through the second hole in order to disturb the motion on the other side and induce interference. Of course, one can always involve the quantum potential as an explanation but then one should explain why this potential exists and the problem is therefore not removed. Many authors thinking about this problem claimed that the guiding wave should carry some energy and the particle should get less and less energy while crossing an interferometer with more and more gates and doors [20]. But, obviously, this is not what is predicted, neither by quantum mechanics nor pilot-wave theory. Another point was that if the empty wave reacts on the particle during the double-slit interference experiment, why should it not also act on some other systems [41]? Could we detect an empty wave? While working on this problem it was realized by L. Hardy [30, 31, 32] that empty waves can sometimes have a physical effect on a second entangled (measuring) system (his idea was actually an adaptation of Elitzur and Vaidman's "interaction free-measurement" protocol [27]) and he found during his research a very fascinating Bell's theorem without inequality involving strange non-local features and questioning the possibility of building up a Lorentz invariant hidden variable model. The result of Hardy is intriguing and also disappointing since, again, it is an indirect effect on hidden variables which is observed. The empty wave affects the dynamics of the second system but one must watch correlations between events to see it (otherwise one could send faster-than-light signals with this non-local protocol). For those already not convinced by the pilot-wave approach this definitely could not help. In a different but related context J. Bell in 1980 [9, pp. 111–116] studied the exotic behavior of Bohmian particles diffracted by a screen and interacting with a complex detecting "which-path" device. It was shown that the path followed by the particle is sometimes completely surrealistic and can even reach the wrong detector (this is connected to the fact that the Bohmian trajectory cannot cross in the configuration space). However, this cannot affect the interpretation since this is again hidden and impossible to test experimentally. In a subsequent paper by Englert et al. [28], already mentioned ([21] and [36]), it was shown that the problem is deeper than Bell thought at first, and that this surrealism exists even with simple particles interacting with Stern–Gerlach devices [40]. Therefore, to quote the authors: "the reality attributed to Bohm trajectories is not physical, it is metaphysical" [28]. Lev Vaidman [43] wrote a very pedagogical paper provocatively titled: "The reality in Bohmian quantum mechanics or can you kill with an empty wave bullet?" In his paper, Vaidman explained with very symptomatic and illustrative examples (such as slow bubble traces developing after the passage of the particle even when the particle is not here but elsewhere) that if one is living in a pilot-wave world then entanglement with meters and environment will break all your convictions about causality and localization (i.e. in agreement with Hardy's conclusions).

His paper reviewing the argument presented in [5] showed also that if the empty waves are involved in all these processes then one can actually measure an empty wave function without the particle being present. This relies on protective measurements of position which allow a measure of the wave function density $|\psi(x)|^2$ of the particle at x even if the pilot-wave trajectory never crosses the interaction region centered on x. The concept of protective measurement is a beautiful idea which was introduced by Y. Aharonov and L. Vaidman [1, 4] (see also [2] and [43]). The principle relies on the possibility of adiabatically coupling the measuring device M with the subsystem S in such a way as to induce no significant change in the $|S(0)\rangle$ initial state while disturbing the meter state $|M(0)\rangle$ in an observable fashion. In such an approach, the system S is therefore protected and it is easily shown that one can use this kind of protocol to record information on some local observable such as $|\psi(x)|^2$ or $\mathbf{J}(x)$. The specific example considered in [5] is based, once more, on the infinite potential well but now with a very local interaction with a meter at one point (i.e. $0 < x = x_0 < L$) of the cavity. The total Hamiltonian is

$$\hat{H}(t) = \frac{-\hbar^2}{2m} \frac{\partial^2}{\partial x^2} + \frac{-\hbar^2}{2M} \frac{\partial^2}{\partial X^2} - \hbar\epsilon g(t)\delta(x - x_0)X, \qquad (12.12)$$

where x is the coordinate of the particle of mass m in the box while X is the coordinate of the meter with mass $M \gg m$. The coupling is monitored by the external parameter $g(t)$ such that $\int_{-\infty}^{+\infty} dt g(t) = 1$. If $g(t)$ changes very fast one goes back to the von Neumann regime but here $g(t)$ changes very slowly, i.e., adiabatically, and it vanishes outside the interval $[-T/2, +T/2]$ where it has the typical value $g(t) \simeq 1/T$. The most characteristic feature of this interaction is of course the presence of the Dirac function which implies a short-range coupling existing only in the vicinity of $x = x_0$. In order to solve the dynamical equation we apply here the adiabatic approximation method [10] and we first search for eigenstates of the equation $\hat{H}(t)\Psi(x, X, t) = E(t)\Psi(x, X, t)$. Inserting the "ansatz" $\Psi(x, X, t) = \phi_s(x, X, t)e^{iPX/\hbar}/\sqrt{(2\pi\hbar)}$ we get the new equation:

$$\left[E(t) - \frac{P^2}{2M} \right] \phi_s(x, X, t) = \frac{-\hbar^2}{2m}\frac{\partial^2}{\partial x^2}\phi_s(x, X, t) - \hbar\epsilon g(t)\delta(x - x_0)X\phi_s(x, X, t).$$

(12.13)

This is actually a 1D Green function problem with t and X as parameters, and Aharonov et al. solved it analytically [5]. Still, since we suppose the coupling to be weak we can alternatively use (as they did as well) the first-order perturbation approximation which leads to $\phi_s(x, X, t) \simeq \phi_n(x) = \sqrt{\frac{2}{L}} \sin(n\pi x/L)$ and $E_{n,P}(X, t) - P^2/2M = (\hbar n\pi/L)^2/(2m) + \delta E$ with

$$\delta E = -\hbar\epsilon g(t)X\langle n|\delta(\hat{x} - x_0)|n\rangle = -\hbar\epsilon g(t)X|\phi_n(x_0)|^2.$$

(12.14)

We point out that there is actually a small slope discontinuity at x_0 since for the 1D Green function we must have

$$d\phi_s(x, X, t)/dx|_{x_0+\delta} - d\phi_s(x, X, t)/dx|_{x_0-\delta}$$
$$= -2m\epsilon g(t)X\phi_s(x_0, X, t)/\hbar$$

(12.15)

with $\delta \to 0^+$. In the weak coupling regime we can neglect this effect and therefore the cavity mode can fairly be considered as "protected". The next step is to expand the full system wave function by solving the Schrödinger equation $i\hbar d\Psi(t)/dt = H(t)\Psi(t)$ and using these eigenmodes labeled by the index n of the cavity mode (here we will limit our analysis to $n = 1$) and P the "orthodox" momentum of the pointer. We have

$$\Psi(x, X, t) = \Sigma_{n,P}b_{n,P}(t, X)\Psi_{n,P}(x, X, t).$$

(12.16)

In the adiabatic approximation we write the amplitude coefficients as

$$b_{n,P}(t, X) = c_{n,P}(t, X)e^{-i\int_{-\infty}^{t} dt' E_{n,P}(X, t')/\hbar}$$

(12.17)

and we get here:

$$\Psi(x, X, t) \simeq \phi_s(x, X, t) \int \frac{\mathrm{d}P}{\sqrt{(2\pi\hbar)}} M(P) \mathrm{e}^{\mathrm{i}PX/\hbar} \mathrm{e}^{-\mathrm{i}\frac{P^2}{2M\hbar}t}$$
$$\cdot \, \mathrm{e}^{\mathrm{i}\int_{-\infty}^{t} \mathrm{d}t'[\epsilon g(t')X|\phi_n(x_0)|^2 - \beta_n(t')]} \mathrm{e}^{-\mathrm{i}E_n t/\hbar} \tag{12.18}$$

with $\mathrm{i}\beta_n(t) = \langle n|\frac{\mathrm{d}}{\mathrm{d}t}|n\rangle \simeq 0$. In doing this calculation we supposed that the initial state begins unentangled, i.e., as for the von Neumann procedure. This initial state corresponds to the product of an undisturbed cavity mode $n = 1$ (i.e. $\phi_s(x, X, t) \simeq \sqrt{\frac{2}{L}} \sin(\pi x/L)$) by a localized wave packet with Gaussian Fourier coefficient $M(P) \propto \mathrm{e}^{-\Delta p^2}$. The coupling is supposed to be slow and weak so that the energy given by the interaction is not large enough to induce transition between different eigenmodes. For $t \to +\infty$ we thus get

$$\Psi(x, X, t) \simeq \phi_s(x, X, t) \mathrm{e}^{-\mathrm{i}E_n t/\hbar} \cdot \Psi_M(X, t) \cdot \mathrm{e}^{\mathrm{i}\epsilon X|\phi_n(x_0)|^2}, \tag{12.19}$$

which shows that the main result of the interaction is to induce a phase kick to the pointer wave packet $\Psi_M(X, t)$. If we neglect the free space spreading of the pointer wave packet this phase shift will impose a translation $\Delta P \simeq +\hbar\epsilon|\phi_n(x_0)|^2$ in the Fourier space such as $M'(P) = M(P - \Delta P)$. This results in a protective measurement where the local adiabatic coupling keeps the confined mode $\phi_s(x, X, t)$ undisturbed.

Now comes the paradox: since the cavity mode is protected and since it corresponds to a de Broglie vanishing velocity of the particle S (i.e., $\mathrm{d}x(t)/\mathrm{d}t = \partial_x S(\mathbf{x}, t)/m = 0$) we deduce that the pointer M is disturbed by the local interaction centered at $x = x_0$ even though S never approaches this position. How could that be? For Aharonov et al. one can hardly avoid the conclusion that Bohmian trajectories are just a mathematical construct. The same conclusion was actually given (although in a less technical way) in a previous paper [3] where the authors concluded that the Bohmian trajectory contradicts the natural statement: "an empty wave should not yield observable effects on other particles". Indeed, the measuring device recording $|\psi_n(x_0)|^2$ in the "empty" region surrounding x_0 yields non-null outcomes (identical conclusions were discussed in [6]). In his review paper [43], Vaidman, however, considered the problem from a wider perspective and commented that for him in the framework of Bohmian mechanics there is no fundamental problem since "these experiments are *not* good verification measurements" so that Bohmian proponents have "a good defense". Nevertheless, this looks mysterious or magical since one would like to find where the force acting on the pointer comes from. Furthermore, even if one is not accepting the ontology proposed by de Broglie's pilot wave it was at least possible until now to accept its self-consistency. Does protective measurement change the rules? Indeed, magical forces have no place in physics. In order to remove some of these ambiguities and magical features I developed in a paper published in 2005 [22] a dynamical analysis of the protective measurement discussed in [5] seen from the point of view of pilot-wave theory. I will now summarize my reasoning using the calculations given before. First, we observe that the quantum potential for the system given by Eq. (12.19) is:

$$
Q(x, X, t) = \frac{-\hbar^2 \frac{\partial^2}{\partial x^2}|\Psi(x, X, t)|}{2m} \frac{}{|\Psi(x, X, t)|} + \frac{-\hbar^2 \frac{\partial^2}{\partial X^2}|\Psi(x, X, t)|}{2M} \frac{}{|\Psi(x, X, t)|}
$$

$$
\simeq \frac{-\hbar^2 \frac{\partial^2}{\partial x^2}|\phi_s(x, X, t)|}{2m} \frac{}{|\phi_s(x, X, t)|} + \frac{-\hbar^2 \frac{\partial^2}{\partial X^2}|\Psi_M(X, t)|}{2M} \frac{}{|\Psi_M(X, t)|}. \qquad (12.20)
$$

Here, we fairly neglected the small contributions of terms containing the X derivatives of $|\phi_s(x, X, t)|$. Now, using Eqs. (12.13), (12.14) and the fact that $\phi_s(x, X, t) \simeq \phi_1(x)$ is real we immediately get

$$
Q(x, X, t) \simeq \frac{(\hbar\pi/L)^2}{2m} - \hbar\epsilon g(t)X|\phi_1(x_0)|^2
$$

$$
+ \hbar\epsilon g(t)\delta(x - x_0)X + \frac{-\hbar^2 \frac{\partial^2}{\partial X^2}|\Psi_M(X, t)|}{2M} \frac{}{|\Psi_M(X, t)|}. \qquad (12.21)
$$

Now, the potential acting in the Hamilton–Jacobi equation is $U = V + Q$, where $V = -\hbar\epsilon g(t)\delta(x - x_0)X$ is the "classical" local interaction potential associated with the protective measurement protocol. Here, this leads therefore to

$$
U(x, X, t) \simeq \frac{(\hbar\pi/L)^2}{2m} - \hbar\epsilon g(t)X|\phi_1(x_0)|^2 + \frac{-\hbar^2 \frac{\partial^2}{\partial X^2}|\Psi_M(X, t)|}{2M} \frac{}{|\Psi_M(X, t)|}. \qquad (12.22)
$$

Remarkably, the local potential has been removed from the total Hamiltonian because the singular term in V exactly compensates the one in Q. This implies that from the framework of pilot-wave theory the interaction is highly quantum-like, i.e., it has no classical analogue. This is even more clear in the Newton picture. Newton's law reads indeed $md^2x(t)/dt^2 = F_x$ and $md^2X(t)/dt^2 = F_X$ and with the definition for U this implies for the evolution of S

$$
F_x = -\frac{\partial}{\partial x}U(x, X, t) \simeq 0, \qquad (12.23)
$$

i.e. the force applied on the Bohmian particle vanishes. This situation is exactly similar to the one obtained in the Einstein example or in the s-state atom discussed by de Broglie, Pauli and Einstein. In each case the quantum potential is constant over the region of interest so that the particle can indeed stay in static equilibrium in full agreement with the de Broglie guidance condition $mdx(t)/dt = \partial_x S(x, X, t) = 0$. The big difference is that in the protective measurement there is actually a local force $-\frac{\partial}{\partial x}V$ but its effect is compensated by an additional quantum term in $-\frac{\partial}{\partial x}Q$. Remarkably, the situation is completely different for the meter M since we get

$$F_X \simeq -\frac{\partial}{\partial X}U(x, X, t) \simeq +\hbar\epsilon g(t)|\phi_1(x_0)|^2, \qquad (12.24)$$

in agreement with the momentum kick $\Delta P = \int dt' F_X(t') = +\hbar\epsilon|\phi_1(x_0)|^2$ introduced previously. Therefore, the pointer deviation is completely justified from the point of view of the de Broglie and Bohm approach. However, here the force applied on M is of quantum origin and not the local and classical term $-\frac{\partial}{\partial x}V$.

12.4 Conclusion

Finally, what can we deduce from this story? We reviewed pilot-wave theory and showed that the surrealism objection is very old and goes back to the origin of the theory. Einstein did not like this theory in part because the trajectories predicted in general don't follow our classical intuitions about dynamics. Later, this surrealism was criticized because very often even causality is affected by pilot-wave theory. This of course included non-locality as studied by Bell but also modifications of our intuitions about what a trajectory in an interferometer should be. The work by Aharonov et al. on protective measurements follows this strategy, and indeed, it confirms that pilot-wave theory is not classical. Still, this theory is the only known quantum ontology (Lev Vaidman will certainly not agree here) which is completely self-consistent at the mathematical level and at the same time explains every experimental fact (too many words could be said here about Everett's interpretation [8] and its problems associated with probabilities and this will be therefore omitted). Of course, it is probably only a temporary expedient and pilot-wave theory has no convincing or univocal relativistic generalization, but to quote Bell: "Should it not be taught, not as the only way, but as an antidote to the prevailing complacency?" [see 9, p. 160].

The author thanks Serge Huant for helpful suggestions during the preparation of the manuscript of this chapter.

References

[1] Y. Aharonov and L. Vaidman. Measurement of the Schrödinger wave of a single particle. *Phys. Lett. A*, **178**: 38–42, 1993.

[2] Y. Aharonov and L. Vaidman. Protective measurements. *Ann. NY Acad. Sci. Fundamental Problems in Quantum Theory*, **755**: 361–373, 1995.

[3] Y. Aharonov and L. Vaidman. About position measurements which do not show the Bohmian particle position. In *Bohmian Mechanics and Quantum Theory: an Appraisal*, Cushing, J. T., Fine, A. and Goldstein, S. (eds) Kluwer (Dordrecht), 1996.

[4] Y. Aharonov, J. Anandan, and L. Vaidman. Meaning of the wave function. *Phys. Rev. A*, **47**: 4616–4626, 1993.

[5] Y. Aharonov, B.-G. Englert, and M. O. Scully. Protective measurements and Bohm trajectories. *Phys. Lett. A*, **263**: 137–146, 1999.

[6] Y. Aharonov, N. Erez, and M. O. Scully. Time and ensemble averages in Bohmian mechanics. *Phys. Scr.*, **69**: 81–83, 2004.

[7] G. Bacciagaluppi and A. Valentini. *Quantum Theory at the Crossroads. Reconsidering the 1927 Solvay Conference*. Cambridge University Press (Cambridge), 2009.

[8] J. A. Barrett and P. Byrne. *The Everett Interpretation of Quantum Mechanics. Collected Works 1955–1980 with Commentary*. Princeton University Press (Princeton), 2012.

[9] J. Bell. *Speakable and Unspeakable in Quantum Mechanics*. Cambridge University Press (Cambridge), 1987.

[10] D. Bohm. *Quantum Theory*. Prentice Hall (New York), 1951.

[11] D. Bohm. A suggested interpretation of the quantum theory in terms of 'hidden' variables. I. *Phys. Rev.*, **85**: 166–179, 1952.

[12] D. Bohm. A suggested interpretation of the quantum theory in terms of 'hidden' variables. II. *Phys. Rev.*, **85**: 180–193, 1952.

[13] D. Bohm. Proof that probability density approaches $|\psi|^2$ in causal interpretation of the quantum theory. *Phys. Rev.*, **89**: 458–466, 1953.

[14] D. Bohm and B. J. Hiley. *The Undivided Universe*. Routledge (New York), 1993.

[15] D. Bohm and J. P. Vigier. Model of the causal intepretation of quantum theory in terms of a fluid with irregular fluctuations. *Phys. Rev.*, **96**: 208–216, 1954.

[16] L. de Broglie. Sur la fréquence propre de l'électron. *C. R. Acad. Sci. (Paris)*, **180**: 498–500, 1925.

[17] L. de Broglie. Recherches sur la théorie des quanta (researches on the quantum theory), thesis, Paris, 1924, *Ann. de Physique (Paris)*, **10** (3): 22, 1925.

[18] L. de Broglie. *Ondes et Mouvements*. Gauthier Villars (Paris), 1926.

[19] L. de Broglie. La mécanique ondulatoire et la structure atomique de la matière et du rayonnement. *J. Phys. Radium*, **8**: 225–241, 1927.

[20] L. de Broglie. *Une tantative d'interprétation cauale et non linéaire de la mécanique ondulatoire*. Gauthier Villars (Paris), 1956.

[21] C. Dewdney, L. Hardy, and E. J. Squires. How late measurements of quantum trajectories can fool a detector. *Phys. Lett. A*, **184**: 6–11, 1993.

[22] A. Drezet. Comment on: 'Protective measurements and Bohm trajectories' [*Phys. Lett. A* 263 (1999) 137]. *Phys. Lett. A*, **350**: 416–418, 2005.

[23] D. Dürr, Goldstein S., and N. Zanghì. Quantum equilibrium and the origin of absolute uncertainty. *J. Stat. Phys.*, **67**: 843–907, 1992.

[24] A. Einstein. Über entwicklung unserer Anschauungen über das Wesen und die Konstitution der Strahlung. *Physikalische Zeitschrift*, **10**: 817–825, 1909.

[25] A. Einstein. Elementary considerations on the interpretation of the foundations of quantum mechanics. In *Scientific Papers Presented to Max Born on his Retirement from the Tait Chair of Natural Philosophy in the University of Edinburgh*. Oliver and Boyd (London), 1953.

[26] A. Einstein, B. Podolsky, and N. Rosen. Can the quantum-mechanical description of physical reality be considered complete? *Phys. Rev.*, **4**: 777–780, 1935.

[27] A. C. Elitzur and L. Vaidman. Quantum mechanical interaction-free measurements. *Found. Phys.*, **23**: 987–997, 1993.

[28] B.-G. Englert, M. O. Scully, G. Suüssmann, and H. Walther. Surrealistic Bohm trajectories. *Z. Naturforsch.*, **47a**: 1175–1186, 1992.

[29] S. Gao. Protective measurement and the meaning of the wave function. philsci-archive.pitt.edu/8741/, 2011.

[30] L. Hardy. On the existence of empty waves in quantum theory. *Phys. Lett. A*, **167**: 11–16, 1992.

[31] L. Hardy. Quantum mechanics, local realistic theories, and Lorentz-invariant realistic theories. *Phys. Rev. Lett.*, **68**: 2981–2984, 1992.

[32] L. Hardy and E. J. Squires. On the violation of Lorentz-invariance in the deterministic hidden-variable interpretation of quantum theory. *Phys. Lett. A*, **168**: 169–173, 1992.

[33] W. Heisenberg. *Physics and Philosophy: the Revolution in Modern Physics*. Harper and Row (New York), 1962.

[34] P. R. Holland. *The Quantum Theory of Motion: an Account of the de Broglie–Bohm Causal Interpretation of Quantum Mechanics*. Cambridge University Press (Cambridge), 1993.

[35] E. Madelung. Quantentheorie in hydrodynamischer Form. *Z. Phys.*, **40**: 322–326, 1927.

[36] G. Naaman-Marom, N. Erez, and L. Vaidman. Position measurements in the de Broglie–Bohm interpretation of quantum mechanics. *Ann. Phys.*, **3327**: 2522–2542, 2012.

[37] E. Nelson. Derivation of the Schrödinger equation from Newtonian mechanics. *Phys. Rev.*, **150**: 1079–1085, 1966.

[38] W. Pauli. Remarques sur le problème des parameters caches dans la mécanîque quantique et sur la théorie de l'orde pilote. In A. George (ed.), *Louis de Broglie: Physicien et Penseur*, pp. 33–42. Albin Michel (Paris), 1953.

[39] N. Rosen. On waves and particles. *J. of the Elisha Mitchell Scientific Soc.*, **61**: 67–73, 1945.

[40] M. O. Scully. Do Bohm trajectories always provide a trustworthy physical picture of particle motion? *Phys. Scr.*, **T76**: 41–46, 1998.

[41] F. Selleri. *Quantum Paradoxes and Physical Reality*. Kluwer Academic Publishers (Dordrecht), 1990.

[42] Institut Solvay (ed.) *Electrons et Photons: Rapport du cinquième congrès Solvay, 1927*. Gauthier Villars (Paris), 1928.

[43] L. Vaidman. The reality in Bohmian quantum mechanics or can you kill with an empty wave bullet? *Found. Phys.*, **35**: 299–312, 2005.

[44] A. Valentini. Signal-locality, uncertainty and the subquantum H theorem, I. *Phys. Lett. A*, **156**: 5–11, 1991.

[45] A. Valentini. Signal-locality, uncertainty and the subquantum H theorem, II. *Phys. Lett. A*, **158**: 1–8, 1991.

13

Entanglement, scaling, and the meaning of the wave function in protective measurement

MAXIMILIAN SCHLOSSHAUER AND TANGEREEN V. B. CLARINGBOLD

We examine the entanglement and state disturbance arising in a protective measurement and argue that these inescapable effects doom the claim that protective measurement establishes the reality of the wave function. An additional challenge to this claim results from the exponential number of protective measurements required to reconstruct multi-qubit states. We suggest that the failure of protective measurement to settle the question of the meaning of the wave function is entirely expected, for protective measurement is but an application of the standard quantum formalism, and none of the hard foundational questions can ever be settled in this way.

13.1 Introduction

From the start, the technical result of protective measurement has been suggested to have implications for the interpretation of quantum mechanics. Consider how Aharonov and Vaidman [2] chose to begin their original paper introducing the idea of protective measurement:

We show that it is possible to measure the Schrödinger wave of a single quantum system. This provides a strong argument for associating physical reality with the quantum state of a single system . . . .

Since then, the pioneers of protective measurement seem to have taken a more moderate stance. Vaidman [42], in a recent synopsis of protective measurement, concedes that

the protective measurement procedure is not a proof that we should adopt one interpretation instead of the other, but it is a good testbed which shows advantages and disadvantages of various interpretations.

Notwithstanding this more subtle perspective and a number of critical studies of the technical and foundational aspects of protective measurement,[1] Gao [21] has maintained, if not amplified, the force of Aharonov and Vaidman's original argument:

An immediate implication is that the result of a protective measurement, namely the expectation value of the measured observable in the measured state, reflects the actual physical property of the measured system, as the system is not disturbed after this result has been obtained. ... Moreover, since the wave function can be reconstructed from the expectation values of a sufficient number of observables, the wave function of a quantum system is a representation of the physical state (or ontic state) of the system.

Clearly, if we could reliably measure the unknown quantum state of a single quantum system without changing that state, it would be entirely sensible – and perhaps even inevitable – to admit the objective, physical reality of this state. Such a measurement, however, is impossible, and no measurement scheme based on an application of the standard quantum formalism, protective measurement included, can rise above this intrinsic limitation [12].

It follows that whatever form the "measurement of the wave function" takes in protective measurement, it must be weaker than the condition we stated in the previous paragraph: "If we could *reliably* measure the *unknown* quantum state of a single quantum system *without changing* that state ... ." The italicized words indicate possibilities for relaxing this condition. We might be content with measurements that are not 100% reliable and may change the state, as long as the disturbance can be made arbitrarily small (or unlikely). Or we might be able to show that the measurement is possible only for certain quantum states, or under certain conditions, or both.

Indeed, all of these concessions must be made in the case of protective measurement [3, 13; see also 38, 40]. Of these, the fact that protective measurement only works under carefully designed conditions and for special quantum states – specifically, the system must be in a non-degenerate eigenstate of its Hamiltonian – may well be of least concern. After all, if protective measurement allowed us to operationally establish the reality of an unknown quantum state in certain situations, perhaps it would not be so far-fetched to extend this interpretation to the rest of the states. The more serious issue, however, arises from the inevitable system–apparatus entanglement in protective measurement. This entanglement introduces an irreducible randomness into the readout; there is a non-zero probability for the system to end up in a state different from the initial state. While this issue has been pointed out before [6, 13], here we will take it up in more detail, by describing the creation of entanglement in protective measurement (Section 13.2) and discussing

---

[1] See, for example, Schwinger [35], Rovelli [28], Samuel and Nityananda [29], Unruh [41], Dass and Qureshi [13], Alter and Yamamoto [6], Uffink [38, 40].

the implications for the claim that protective measurement suggests the reality of the wave function (Section 13.3). In Section 13.4, we identify another, more subtle challenge to this claim, namely, the exponential scaling of the number of protective measurements required to reconstruct multi-qubit states. We end on a broader note (Section 13.5), arguing that since protective measurement is an application of the quantum formalism, it cannot settle significant foundational questions.

13.2 Theory of entanglement in protective measurement

Here we will go beyond the zeroth-order limit $T \to \infty$ usually considered in protective measurement [2, 3, 13] and derive an expression for the final system–apparatus state to first order in $1/T$.

Consider two systems S and A described by self-Hamiltonians H_S and H_A, respectively. System A plays the role of a measuring apparatus for S, in the sense that we let S and A interact such that information about S can be transferred to A. The interaction is generated by the interaction Hamiltonian

$$\hat{H}_{\mathrm{I}}(t) = g(t)\hat{P} \otimes \hat{O}, \tag{13.1}$$

where $\hat{P}$ is the momentum operator of A (the canonical conjugate to the position operator $\hat{X}$), and $\hat{O}$ is an arbitrary observable of the system S. The function $g(t)$, with $\int_0^T g(t)\,\mathrm{d}t = 1$, describes the time-dependent strength of the interaction, with $g(t) = 0$ for $t < 0$ and $t > T$. Thus, T describes the duration of the measurement interaction. In contrast with a standard impulsive (strong, von Neumann) measurement, here T is taken to be very large. The normalization condition $\int_0^T g(t)\,\mathrm{d}t = 1$ then implies that the magnitude of $g(t)$ will be small. This results in a weak adiabatic coupling between the system and the apparatus. Neglecting the switching-on and switching-off periods around $t = 0$ and $t = T$ and assuming $g(t)$ remains approximately constant for $t \in [0, T]$, we can write $g(t) = 1/T$. Thus, the total Hamiltonian can be treated as time-independent for the duration of the measurement interaction,

$$\hat{H} = \hat{H}_S + \hat{H}_A + \frac{1}{T}\left(\hat{P} \otimes \hat{O}\right). \tag{13.2}$$

To simplify the formal treatment from here on, let us assume that $\hat{H}_A$ commutes with $\hat{P}$.[2] Then we can find a set of simultaneous eigenstates $\{|A_i\rangle\}$ of $\hat{H}_A$ and $\hat{P}$ such that

$$\hat{H}_A |A_i\rangle = E_i^A |A_i\rangle, \qquad \hat{P}|A_i\rangle = a_i |A_i\rangle. \tag{13.3}$$

---

[2] This assumption is not necessary for a protective measurement to obtain [13].

Therefore, the exact eigenstates of the full Hamiltonian $\hat{H}$ can be written in product form as $\left| E_m^S(a_i) \right\rangle |A_i\rangle$, where the $\left| E_m^S(a_i) \right\rangle$ are the eigenstates of the system part of $\hat{H}$, which is given by

$$\hat{H}_S'(a_i) = \hat{H}_S + \frac{1}{T} a_i \hat{O}. \tag{13.4}$$

Note that $\hat{H}_S'(a_i)$, and thus also its eigenstates $\left| E_m^S(a_i) \right\rangle$, explicitly depend on the eigenvalue a_i of $\hat{P}$.

Suppose now that S is initially in a non-degenerate eigenstate $|n\rangle$ of $\hat{H}_S$ (but not necessarily of $\hat{O}$)[3] with eigenvalue E_n, and let the pointer of A be described by a Gaussian wave packet $|\phi(x_0)\rangle$ of eigenstates of $\hat{X}$ centered around x_0. Thus, the initial composite state of system and apparatus is

$$|\Psi(t = 0)\rangle = |n\rangle \, |\phi(x_0)\rangle. \tag{13.5}$$

Since $\hat{H}$ is time-independent, at $t = T$ this state has evolved into (taking $\hbar \equiv 1$)

$$|\Psi(t = T)\rangle = e^{-i\hat{H}T} |n\rangle \, |\phi(x_0)\rangle. \tag{13.6}$$

Inserting a complete set of eigenstates $\left| E_m^S(a_i) \right\rangle |A_i\rangle$ of H, we obtain

$$\begin{aligned}
|\Psi(t = T)\rangle &= e^{-i\hat{H}T} \left(\sum_{m,i} \left| E_m^S(a_i) \right\rangle |A_i\rangle \, \langle A_i| \langle E_m^S(a_i)| \right) |n\rangle \, |\phi(x_0)\rangle \\
&= e^{-i\hat{H}T} \sum_{m,i} \left(\langle E_m^S(a_i)|n\rangle \langle A_i|\phi(x_0)\rangle \right) \left| E_m^S(a_i) \right\rangle |A_i\rangle \\
&= \sum_{m,i} e^{-iE(m,a_i)T} \left(\langle E_m^S(a_i)|n\rangle \langle A_i|\phi(x_0)\rangle \right) \left| E_m^S(a_i) \right\rangle |A_i\rangle, \tag{13.7}
\end{aligned}$$

where

$$E(m, a_i) = E_i^A + \frac{1}{T} a_i \langle E_m^S(a_i)|\hat{O} \left| E_m^S(a_i) \right\rangle + \langle E_m^S(a_i)|\hat{H}_S \left| E_m^S(a_i) \right\rangle \tag{13.8}$$

are the eigenvalues of $\hat{H}$ corresponding to the states $\left| E_m^S(a_i) \right\rangle |A_i\rangle$.

By regarding $\hat{H}_I$ as a perturbation to $\hat{H}_S + \hat{H}_A$, we can write down the perturbative expansion of the exact eigenstates $\left| E_m^S(a_i) \right\rangle |A_i\rangle$ of $\hat{H}$,

$$\left| E_m^S(a_i) \right\rangle |A_i\rangle = |m\rangle \, |A_i\rangle + \frac{1}{T} \left(\sum_{m' \neq m} \frac{\langle m'|\hat{O}|m\rangle}{E_m - E_{m'}} |m'\rangle \right) a_i \, |A_i\rangle + O(1/T^2). \tag{13.9}$$

In the limit $T \rightarrow \infty$ usually considered in the treatment of protective measurement, we can therefore replace all states $\left| E_m^S(a_i) \right\rangle$ in Eq. (13.7) by the unperturbed

---

[3] See the discussion by Uffink [38, 40] and Gao [20].

eigenstates $|m\rangle$ of $\hat{H}_S$. Reintroducing the operators $\hat{H}_A$ and $\hat{P}$ in the exponent of the time-evolution operator, this results in the non-entangled final state

$$
\begin{aligned}
|\Psi(t = T)\rangle &= \sum_i e^{-i\hat{H}_A T - iE_n T - i\hat{P}\langle n|\hat{O}|n\rangle} \langle A_i|\phi(x_0)\rangle |n\rangle |A_i\rangle \\
&= e^{-iE_n T} |n\rangle e^{-i\hat{H}_A T} e^{-i\hat{P}\langle n|\hat{O}|n\rangle} \sum_i \langle A_i|\phi(x_0)\rangle |n\rangle |A_i\rangle \\
&= e^{-iE_n T} |n\rangle e^{-i\hat{H}_A T} e^{-i\hat{P}\langle n|\hat{O}|n\rangle} |\phi(x_0)\rangle .
\end{aligned}
\tag{13.10}
$$

Since $e^{-i\hat{P}\Delta x}$ is the translation operator, the term $e^{-i\hat{P}\langle n|\hat{O}|n\rangle}$ applied to the initial wave packet $|\phi(x_0)\rangle$ will shift the center of the wave packet by an amount equal to $\langle n|\hat{O}|n\rangle \equiv \langle\hat{O}\rangle_n$, which is the expectation value of $\hat{O}$ in the initial state $|n\rangle$ of the system. Thus, to zeroth order, the final system–apparatus state is

$$
|\Psi(t = T)\rangle = e^{-iE_n T} |n\rangle e^{-i\hat{H}_A T} |\phi(x_0 + \langle\hat{O}\rangle_n)\rangle .
\tag{13.11}
$$

This establishes the familiar main result of protective measurement: information about the expectation value of $\hat{O}$ in state $|n\rangle$ has been transferred to the apparatus [2, 3, 13].

The crucial point, however, for our subsequent discussion is the observation that for any finite value of T, the system–apparatus state (13.7) is entangled, and therefore the initial state of the system, $|n\rangle$, has been changed. To explicitly see this, we insert expansion (13.9) into Eq. (13.7). Keeping only terms up to $O(1/T)$ and using the first-order perturbative approximation to the energy eigenvalues $E(m, a_i)$, $E(m, a_i) \approx E_i^A + \frac{1}{T} a_i \langle\hat{O}\rangle_m + E_m$, we find (again reintroducing the operators $\hat{H}_A$ and $\hat{P}$)

$$
|\Psi(t = T)\rangle = e^{-iE_n T} |n\rangle e^{-i\hat{H}_A T} |\phi(x_0 + \langle\hat{O}\rangle_n)\rangle + \frac{1}{T} e^{-i\hat{H}_A T} \sum_{m \neq n} \frac{1}{E_n - E_m}
$$
$$
\times \left[\langle m|\hat{O}|n\rangle e^{-iE_n T} e^{-i\hat{P}\langle\hat{O}\rangle_n} - \langle n|\hat{O}|m\rangle e^{-iE_m T} e^{-i\hat{P}\langle\hat{O}\rangle_m} \right] |m\rangle |\widetilde{\phi}(x_0)\rangle ,
\tag{13.12}
$$

where $|\widetilde{\phi}(x_0)\rangle = \sum_i a_i |A_i\rangle \langle A_i|\phi(x_0)\rangle$ is a distorted version of the initial pointer wave packet $|\phi(x_0)\rangle = \sum_i |A_i\rangle \langle A_i|\phi(x_0)\rangle$. (One may also include the second-order perturbative correction to the energy eigenvalues such that the argument of the time-evolution operator is to first order in $1/T$; this correction, however, is irrelevant to the argument below and will therefore be neglected.) The operator $e^{-i\hat{P}\langle\hat{O}\rangle}$ then shifts $|\widetilde{\phi}(x_0)\rangle$ by an amount $\langle\hat{O}\rangle$, leading to the final state

$$|\Psi(t = T)\rangle = e^{-iE_n T} |n\rangle e^{-i\hat{H}_A T} |\phi(x_0 + \langle \hat{O} \rangle_n)\rangle + \frac{1}{T} e^{-i\hat{H}_A T} \sum_{m \neq n} \frac{1}{E_n - E_m} |m\rangle$$

$$\times \left[\langle m|\hat{O}|n\rangle e^{-iE_n T} |\widetilde{\phi}(x_0 + \langle \hat{O} \rangle_n)\rangle - \langle n|\hat{O}|m\rangle e^{-iE_m T} |\widetilde{\phi}(x_0 + \langle \hat{O} \rangle_m)\rangle \right]. \quad (13.13)$$

The first term on the right-hand side is the familiar zeroth-order term of Eq. (13.11). The second term represents quantum correlations between all other eigenstates $|m\rangle \neq |n\rangle$ of $\hat{H}_S$ and wave packets representing shifted apparatus pointers. Thus, Eq. (13.13) describes an entangled superposition involving all possible energy eigenstates $\{|m\rangle\}$ of the system correlated with pointer states. In particular, we see that a readout of the apparatus pointer will indicate, with probability proportional to $1/T^2$, the expectation value of $\hat{O}$ in a state $|m\rangle$ orthogonal to the initial state $|n\rangle$ of the system, with the system then left in this orthogonal state $|m\rangle$ and not in the initial state $|n\rangle$. We also see that, again with probability proportional to $1/T^2$, the apparatus pointer may indicate the expectation value of $\hat{O}$ in the initial state $|n\rangle$ while the system has been projected onto the orthogonal subspace spanned by $\{|m\rangle\}_{m \neq n}$.

13.3 Implications of entanglement in protective measurement

The finite system–apparatus entanglement arising in a protective measurement entails that protective measurements can never transcend the irreducibly probabilistic element inherent in any measurement of a (fully or partially) unknown quantum state. The problem is not only that the quantum state reconstructed from protective measurements may well be (unpredictably) different from the initial state of the system we had set out to measure. The problem is also that even on those occasions when the protective measurement succeeds – i.e., when the collapsed state of the apparatus pointer indicates the expectation value corresponding to the initial state of the system – we cannot infer from the readout of the pointer that we have indeed obtained information about the initial state of the system, rather than about any other state. This is so because *there is no possibility of knowing whether we have succeeded*: while the final pointer measurement may project the system back onto its initial state, the readout itself cannot tell us whether this has actually happened.

Gao [21] misses this important point when he suggests that "when the measurement obtains the expectation value of the measured observable, the state of the measured system is not disturbed." But there is a crucial difference between a situation in which the system remains in the initial state throughout, and a situation in which the system becomes entangled with the apparatus and, through a secondary measurement and with probability less than one, is subsequently projected back onto the initial state. Only the former situation would permit conclusions about having measured the initial state of the system, while it is the latter situation that applies to protective measurements. Thus, *pace* Gao, the state of the measured system is *always* disturbed in a protective measurement.

This undermines the claim that the reconstructed quantum state must be a real, objective property, in the sense that it must have already existed prior to the measurement [for a similar conclusion, see 5, 6, 13]. Indeed, any measurement worth its name – any measurement that allows us to obtain *new* information about the system – will entangle the system and the apparatus, thus introducing an element of randomness and, in this sense, disturbing the initial state of the system; this is as true for a protective measurement as it is for any other kind of quantum measurement. Whether the measurement is strong, weak, protective, or "reversible" [37, 25]; whether we perform a sequence of measurements on a single system or just one measurement: the maximum possible information gain will always be the same [12].[4]

Elaborating on his claim that protective measurement shows that "the wave function of a quantum system is a representation of the physical state of the system", Gao [21] dismisses concerns about state disturbance by arguing that the probability for collapsing the system's state to an orthogonal outcome "can be made arbitrarily small in principle when T approaches infinity." But, from the foundational point of view relevant here, *any* non-zero value of this probability, no matter how small it may be made, will spoil the claim that protective measurement permits us to learn the initial state of the system and that it thus demonstrates the reality of the wave function. The limit $T \to \infty$ required for a reliable protective measurement can never be attained – and if it *could*, we would only have time, so to speak, to protectively measure one observable, thus precluding the possibility of reconstructing the wave function. It is impossible *in principle* to reliably determine the expectation values required to reconstruct the wave function; it is not just impossible in practice. (We agree with Englert [15] that to call something "possible in principle" is simply meaningless. Either, whatever action is contemplated is *possible in practice*, or it is *impossible in principle*.)

The issue at stake here bears some similarity to the situation in quantum tomography, where a quantum state is reconstructed from projective measurements on an ensemble of identically prepared systems. Of course, in the hypothetical case of infinitely many measurements on an infinitely large ensemble, the state determination would be exact. But it is impossible in principle to generate such an ensemble,

---

[4] Alter and Yamamoto [5] have constructed a scheme for measuring a single squeezed harmonic-oscillator state in such a way that the final system–apparatus state is deterministically returned to a disentangled state. But as the authors themselves point out, implementation of this scheme requires full *a priori* knowledge of the state of the system, which means that no information is gained in such a "measurement" [see also 4, 6].

or to carry out infinitely many measurements. Consider one more related, and this time purely classical, example. It is well known that the concept of probability cannot be derived from relative frequencies, simply because we cannot have an infinite number of trials. For any finite number of trials, all we can say is that it is *unlikely* that the measured relative frequency will deviate much from the probability we have inferred from that frequency. "Unlikely", however, already presumes a notion of probability, making the derivation circular. To reply to this conclusion by saying that one may consider infinitely many trials or infinitely large ensembles "in principle" is to miss the point: we are, after all, aiming at a fundamental definition that connects an abstract concept (probability) with what can be measured (relative frequencies). Infinitely large ensembles simply do not exist, and no matter how large we may make the ensemble, at the end we must make a quantitative, *probabilistic* judgment about the correspondence between the theoretical probability value and the relative frequency value.

The case of protective measurement highlights how vigilant one needs to be when using results obtained from mathematical idealizations to justify conclusions pertaining to fundamental questions of nature and interpretation. Just as in quantum tomography, lack of accuracy may not matter as far as practical implementation is concerned: a method that gives us an approximate picture is often all we need. But in-principle lack of accuracy may become decisive when the very procedure is claimed to have implications for our conceptual understanding of the theory itself. It certainly is decisive in the case of protective measurement, refuting the claim that protective measurement has demonstrated the ontological status of the wave function. It follows that secondary claims based on this claim must fail, too; an example is Gao's [19] suggestion that protective measurement effortlessly establishes a result equivalent to the theorem derived by Pusey et al. [27] – namely, that the wave function must be "uniquely determined by the underlying physical state."[5]

13.4 The scaling problem

We now turn our attention to the question of the number of protective measurements of expectation values required to (approximately) reconstruct a wave function; in particular, we will analyze how this number scales with the size of the system. This, at first glance, may appear to be a question of purely practical concern. However, as we will indicate below, it may have foundational implications as well.

---

[5] It should be mentioned that such a result does not conclusively follow from the theorem of Pusey et al. [27] either; for critical discussions, see Colbeck and Renner [11], Hardy [23], Schlosshauer and Fine [33, 34], Wallden [43].

Consider a qubit described by an arbitrary density matrix $\hat{\rho}$. In the Bloch representation, this density matrix can be written as

$$\hat{\rho} = \frac{1}{2} \left(\mathbb{I} + \mathbf{n} \cdot \hat{\boldsymbol{\sigma}} \right). \tag{13.14}$$

Here, $\mathbb{I}$ denotes the identity operator. The components of the operator $\hat{\boldsymbol{\sigma}}$ are given by the Pauli operators $\hat{\sigma}_x$, $\hat{\sigma}_y$, and $\hat{\sigma}_z$, and the real-valued components of the vector $\mathbf{n}$ are given by the expectation values of $\hat{\sigma}_x$, $\hat{\sigma}_y$, and $\hat{\sigma}_z$ in the state $\hat{\rho}$, i.e.,

$$n_i = \text{Tr}\left[\hat{\sigma}_i \hat{\rho}\right] = \langle \hat{\sigma}_i \rangle_{\hat{\rho}}, \qquad i = x, y, z. \tag{13.15}$$

It follows that the density matrix of a qubit is uniquely determined by the three expectation values $\langle \hat{\sigma}_x \rangle_{\hat{\rho}}$, $\langle \hat{\sigma}_y \rangle_{\hat{\rho}}$, and $\langle \hat{\sigma}_z \rangle_{\hat{\rho}}$. We are free, of course, to choose other triples of observables, as long as these observables form an informationally complete set (i.e., as long as their expectation values uniquely determine the qubit state). But we always need to measure the expectation values of three such observables to reconstruct an arbitary state $\hat{\rho}$; no pair of observables will do.

In the general case of a d-dimensional Hilbert space, the measurement of the expectation values of a minimum of $n(d) = d^2 - 1$ observables will be required to determine an arbitrary state $\hat{\rho}$. Therefore, determining an arbitrary N-qubit state requires at least

$$n(N) = d^2 - 1 = (2^N)^2 - 1 = 4^N - 1 \tag{13.16}$$

expectation values to be measured, which means that the required number of measurements grows exponentially with N. The number of observables can be reduced, however, if prior information about the state $\hat{\rho}$ is available. For example, if $\hat{\rho}$ is known to be pure, then the number of observables required to uniquely determine the state only scales linearly with d [24, 10]. Protective measurement may avail itself to such a reduction in the number of required measurements, since the system has to be in a pure eigenstate of the self-Hamiltonian of the system. Even so, since $d = 2^N$ for an N-qubit state, this still results in an exponential scaling behavior.

Since no concrete experimental realizations of protective measurement are presently available, it is difficult to provide a good estimate of the time that would be required to reconstruct the wave function of a multi-qubit system with a degree of accuracy comparable to that typically achieved in quantum tomography. Dickson [14] points out that the interaction time T may only need to be large on an atomic scale; without giving further details, he provides an estimate of 10^{-5} seconds. But this value seems unduly low in light of the fact that even the impulsive measurements used in quantum tomography often take longer. For example, Häffner et al. [22] have carried out quantum tomography on an eight-qubit state of trapped $^{40}\text{Ca}^+$ ions, requiring 656,100 measurements and a total measurement time of 10 hours, or about 0.05 seconds per measurement.

Whatever estimates for the measurement duration T might be reasonable in potential practical implementations of protective measurement, it is clear that because of the exponential scaling, reconstructions of quantum states of large systems (say, hundreds of qubits) would require astronomically long total measurement times. Of course, this is no different from standard quantum tomography. But contrary to protective measurement, quantum tomography has not been associated with the claim that it demonstrates the reality of the quantum state (the reason being as we would expect the proponent of protective measurement to argue, that quantum tomography works with ensembles, not single systems). Thus, there is a much greater burden on protective measurement to show that its method for state reconstruction has the suggested physical meaning and foundational implications.

To reconstruct a 100-qubit pure state would require the measurement of about $2^{100} \approx 10^{30}$ expectation values, and even if we accept Dickson's optimistic estimate of $T \approx 10^{-5}$ s for a single protective measurement, the time needed to measure such a state would exceed the present age of the Universe by many orders of magnitude. Is this merely a practical problem? We are not sure it is. Clearly, in contrast with the discussion in the preceding section, the issue is no longer about the tension between a strict mathematical limit and the inevitably finite version attainable in practice. Yet, even if we assume that the Universe will continue to exist for a sufficiently long time for such a large number of measurements to be carried out, there is certainly no sense in which this experiment could ever be realized. As Englert [15] put it,

Statements like "In principle, I could solve the Schrödinger equation to predict the next solar eclipse" are empty unless you can do it in practice.

By the same token, statements like "In principle, I could carry out a protective measurement of a 100-qubit state to establish its reality" must be considered empty. But if there is no possibility for this state to be measured, protectively or otherwise, what can such a state possibly mean?[6] We must leave this question open; our aim here has been to point out that even ostensibly mundane practical constraints may have fundamental implications for the question of how and whether protective measurement could decide the question of the ontological meaning of the wave function.

13.5 Protective measurement and the quantum formalism

Protective measurement does not demonstrate the physical reality of the wave function. Should this result be surprising? Did we, and other authors [35, 28, 29, 41, 13, 6, 38, 40], really need to invoke various, ostensibly technical arguments to come to this conclusion?

---

[6] Aaronson [1] has posed a similar question in the context of quantum tomography.

Quantum mechanics provides a formalism for relating and transforming probability assignments concerning outcomes of future measurements. The notion of measurement and the existence of outcomes are all taken to be primitives of the theory. (To repeat a popular analogy, this is just as in classical probability theory, which neither explains the existence of dice and nor why throwing them results in particular results.) To want to say more is to tack onto the quantum formalism a story: of metaphysics, say, or of causation. But as the wealth of competing interpretations of quantum mechanics shows, the choice of any such particular story is hopelessly underdetermined by the quantum formalism itself. The formalism does not mandate any particular ontological commitment toward the interpretation of its elements, quantum states and their corresponding probabilities included.[7] When we calculate probabilities from quantum states, we start from some initial quantum-state assignment. But the question of what this assignment physically *means* or *represents* is of no relevance, because any probabilistic predictions derived from the quantum formalism, using the initial state assignment, are insensitive to how we choose to answer the question. We get the quantum formalism cranking to obtain a new quantum state, and there is no reason to apply to this state an interpretation different from the interpretation we chose to give our original state assignment.

The point here is that no application of the quantum formalism, and no observational data that is in agreement with the predictions of this formalism, can provide definite answers to questions about the interpretation of the quantum state. Protective measurement is just such an application. Therefore, it is not equipped to settle the significant foundational and interpretive questions, no matter how wishful the thinking. (By "significant" we mean the hard questions – the question of the meaning of the wave function, for example – rather than the "softer" questions about the explanatory power or the reasonableness of individual interpretations of quantum mechanics.)

Of course this is not to say that by milking the quantum formalism we cannot produce something fresh. Quantum information theory and decoherence theory are good examples, but they, just like protective measurement, have not answered the hard interpretive questions; and they, too, could not be expected to do so. Quantum information theory may have motivated new information-based interpretations of quantum mechanics, but there are quantum information theorists who are Bohmians and others who are Everettians. Decoherence, it is to be remembered, is an essentially technical result about the dynamics and measurement statistics of open quantum systems. In particular, its predictively relevant part relies on reduced density matrices, whose formalism and interpretation presume the collapse postulate and Born's rule. Thus if we understand the quantum measurement problem as the

---

[7] Indeed, one can construct a picture of quantum mechanics in which quantum states are nothing but a representation of our personal beliefs about our future experiences when we interact with a quantum system [16].

question of how to reconcile the linear, deterministic evolution described by the Schrödinger equation with the occurrence of random, definite measurement outcomes, then decoherence has certainly not solved this problem, as is now widely recognized [30, 31]. What decoherence rather solves is a *consistency problem*: the problem of explaining why and when quantum probability distributions approach the classically expected distributions. But this is a purely practical problem, not a game-changer for quantum foundations. To be sure, the picture associated with the decoherence process has sometimes been claimed to be suggestive of particular interpretations of quantum mechanics[8] or to pinpoint internal concistency issues [30]. But it might be safer to say that certain interpretations (such as the Everett interpretation) are simply more *in need* of decoherence to define their structure. At the end of the day, any interpretation that does not involve entities, claims, or structures in contradiction with the prediction of decoherence theory (which is to say, with the predictions of quantum mechanics) will remain viable.

It follows that if we hope to make headway in foundational matters, we have to consider theories beyond quantum mechanics and study how their predictions match those of quantum mechanics. Reconstructions of quantum mechanics are one example of this approach; they have shown that features traditionally regarded as uniquely quantum – such as interference, Bell-type violations, no-signaling and no-cloning constraints, and state disturbance through measurement – are generic to entire classes of probabilistic theories. Another example is Bell's theorem [7, 8], although what exactly the experimentally measured violations of Bell's inequalities tell us about nature remains a matter of debate [32]. Like Bell's theorem, the PBR theorem [27] is based on the consideration of hidden-variables models and accommodates a variety of conclusions [11, 23, 33, 34]. Thus, a decisive answer to a foundational question may elude us even if we consider models beyond quantum mechanics.

13.6 Concluding remarks

In response to Uffink's [38] criticism of protective measurement, Gao [18] writes:

It seems that the errors in Uffink's arguments were made at least partly due to his biased philosophical opinions. Why protect the interpretation of the wave function against protective measurements? Why make the different views on the meaning of the wave function peacefully coexist? Is it not very exciting and satisfying if we can decide the issue of the interpretation of the wave function someday? Is it not one of the ultimate objectives of our explorations in quantum foundations?

---

[8] Indeed, historically decoherence theory arose in the context of Zeh's independent formulation of an Everett-style interpretation [44, 9].

To this, Uffink [39] replies:

Of course, [I agree fully] with Gao that such an alternative view would be much more desirable. However, apart from the hot aspirations we might all have concerning the interpretation of quantum theory, we also need the cool breeze of critical analysis before we step forward.

Uffink's attitude, like our own, is not meant to be pessimistic. It merely reflects a realistic assessment of aims and means. Protective measurement is an ingenious implementation of a quantum measurement, but a quantum measurement it nevertheless remains. As such, it simply cannot, even in principle, accurately determine the wave function of a single system. In particular, we have pointed to two problems: the necessarily finite interaction time and the astronomically large number of measurements required for bigger systems. The probabilistic, random element of any quantum measurement remains; there cannot be any information gain without disturbance. But only if perfectly reliable, non-disturbing state determination were possible would protective measurement qualify as an arbiter in the question of the nature of the wave function. As we have argued, the failure of protective measurement to accomplish this goal is not surprising, for no application of the quantum formalism can bypass the fundamental indifference of this formalism to its interpretation.

If protective measurement had indeed established the reality of the wave function (or its direct correspondence with reality), then, without doubt, we would have happily concurred with Gao's [21] assessment of protective measurement as a "paradigm shift in understanding quantum mechanics." As it stands, however, not only do all interpretive options remain on the table, but, in our view, protective measurement also fails to nudge us one way or the other. If one does not already believe in the reality of the wave function, then what does protective measurement offer to change one's mind? Not only does protective measurement fail to challenge the epistemic view of the wave function, but it also leaves untouched all the features that make the epistemic view so attractive and powerful in the first place [36, 17, 26]. To say so is not to diminish the practical usefulness of protective measurement or to discourage its future exploration, but to recognize the fundamental limitations when using the quantum formalism to provide its own interpretation.

References

[1] Aaronson, S. 2007. The learnability of quantum states. *Proc. R. Soc. A*, **463**, 3089–3114.

[2] Aharonov, Y. and Vaidman, L. 1993. Measurement of the Schrödinger wave of a single particle. *Phys. Lett. A*, **178**, 38–42.

[3] Aharonov, Y., Anandan, J., and Vaidman, L. 1993. Meaning of the wave function. *Phys. Rev. A*, **47**, 4616–4626.

[4] Alter, O. and Yamamoto, Y. 1995. Inhibition of the measurement of the wave function of a single quantum system in repeated weak quantum nondemolition measurements. *Phys. Rev. Lett.*, **74**, 4106–4109.

[5] Alter, O. and Yamamoto, Y. 1996. Protective measurement of the wave function of a single squeezed harmonic oscillator state. *Phys. Rev. A*, **53**, R2911–R2914.

[6] Alter, O. and Yamamoto, Y. 1997. Reply to "Comment on 'Protective measurement of the wave function of a single squeezed harmonic oscillator state' ". *Phys. Rev. A*, **56**, 1057–1059.

[7] Bell, J. S. 1964. On the Einstein–Podolsky–Rosen paradox. *Physics*, **1**, 195–200.

[8] Bell, J. S. 1966. On the problem of hidden variables in quantum mechanics. *Rev. Mod. Phys.*, **38**, 447–452.

[9] Camilleri, K. 2009. A history of entanglement: decoherence and the interpretation problem. *Stud. Hist. Phil. Mod. Phys.*, **40**, 290–302.

[10] Chen, J., Dawkins, H., Ji, Z., et al. 2013. Uniqueness of quantum states compatible with given measurement results. *Phys. Rev. A*, **88**, 012109.

[11] Colbeck, R. and Renner, R. 2012. Is a system's wave function in one-to-one correspondence with its elements of reality? *Phys. Rev. Lett.*, **108**, 150402.

[12] D'Ariano, G. M. and Yuen, H. P. 1996. Impossibility of measuring the wave function of a single quantum system. *Phys. Rev. Lett.*, **76**, 2832–2835.

[13] Dass, N. D. H. and Qureshi, T. 1999. Critique of protective measurements. *Phys. Rev. A*, **59**(4), 2590–2601.

[14] Dickson, M. 1995. An empirical reply to empiricism: protective measurement opens the door for quantum realism. *Phil. Sci.*, **62**, 122–140.

[15] Englert, B.-G. 2013. On quantum theory. *Eur. Phys. J. D*, **67**, 238.

[16] Fuchs, C. A. and Schack, R. 2013. Quantum-Bayesian coherence. *Rev. Mod. Phys.*, **85**, 1693–1715.

[17] Fuchs, C. A. 2010. QBism, the perimeter of quantum Bayesianism. arXiv: 1003.5209v1 [quant-ph].

[18] Gao, S. 2011. Comment on "How to protect the interpretation of the wave function against protective measurements" by Jos Uffink. philsci-archive.pitt.edu/8942.

[19] Gao, S. 2013a. Distinct quantum states cannot be compatible with a single state of reality. philsci-archive.pitt.edu/9609.

[20] Gao, S. 2013b. On Uffink's criticism of protective measurements. *Stud. Hist. Phil. Mod. Phys.*, **44**, 513–518.

[21] Gao, S. 2013c. Protective measurement: a paradigm shift in understanding quantum mechanics. philsci-archive.pitt.edu/9627.

[22] Häffner, H., Hänsel, W., Roos, C. F., et al. 2005. Scalable multiparticle entanglement of trapped ions. *Nature*, **438**, 643–646.

[23] Hardy, L. 2012. Are quantum states real? arXiv:1205.1439v3 [quant-ph].

[24] Heinosaari, T., Mazzarella, L., and Wolf, M. M. 2013. Quantum tomography under prior information. *Comm. Math. Phys.*, **318**, 355–374.

[25] İmamoğlu, A. 1993. Logical reversibility in quantum-nondemolition measurements. *Phys. Rev. A*, **47**, R4577–R4580.

[26] Mermin, N. D. 2012. Quantum mechanics: fixing the shifty split. *Phys. Today*, **65**, 8–10.

[27] Pusey, M. F., Barrett, J., and Rudolph, T. 2012. On the reality of the quantum state. *Nature Phys.*, **8**, 475–478.

[28] Rovelli, C. 1994. Comment on "Meaning of the wave function." *Phys. Rev. A*, **50**, 2788–2792.

[29] Samuel, J. and Nityananda, R. 1994. Comment on "Meaning of the wave function." arXiv:gr-qc/9404051v1.

[30] Schlosshauer, M. 2004. Decoherence, the measurement problem, and interpretations of quantum mechanics. *Rev. Mod. Phys.*, **76**, 1267–1305.

[31] Schlosshauer, M. 2007. *Decoherence and the Quantum-to-Classical Transition*. 1st edn. Berlin/Heidelberg: Springer.

[32] Schlosshauer, M. 2011. *Elegance and Enigma: the Quantum Interviews*. 1st edn. Berlin/Heidelberg: Springer.

[33] Schlosshauer, M. and Fine, A. 2012. Implications of the Pusey–Barrett–Rudolph quantum no-go theorem. *Phys. Rev. Lett*, **108**, 260404.

[34] Schlosshauer, M. and Fine, A. 2014. No-go theorem for the composition of quantum systems. *Phys. Rev. Lett.* **112**, 070407

[35] Schwinger, J. 1993. Quantum mechanics: not mysterious. *Science*, **262**, 826–827.

[36] Spekkens, R. W. 2007. Evidence for the epistemic view of quantum states: a toy theory. *Phys. Rev. A*, **75**, 032110.

[37] Ueda, M. and Kitagawa, M. 1992. Reversibility in quantum measurement processes. *Phys. Rev. Lett.*, **68**, 3424–3427.

[38] Uffink, J. 1999. How to protect the interpretation of the wave function against protective measurements. *Phys. Rev. A*, **60**, 3474–3481.

[39] Uffink, J. 2012. Reply to Gao's "Comment on 'How to protect the interpretation of the wave function against protective measurements'". philsci-archive.pitt.edu/9286.

[40] Uffink, J. 2013. Reply to Gao's "On Uffink's criticism of protective measurements." *Stud. Hist. Phil. Mod. Phys.*, **44**, 519–523.

[41] Unruh, W. G. 1994. Reality and measurement of the wave function. *Phys. Rev. A*, **50**, 882–887.

[42] Vaidman, L. 2009. Protective measurements. In: Greenberger, D., Hentschel, K., and Weinert, F. (eds.), *Compendium of Quantum Physics: Concepts, Experiments, History and Philosophy*. Berlin/Heidelberg: Springer, pp. 505–508.

[43] Wallden, P. 2013. Distinguishing initial state-vectors from each other in histories formulations and the PBR argument. *Found. Phys.*, **43**, 1502–1525.

[44] Zeh, H. D. 1970. On the interpretation of measurement in quantum theory. *Found. Phys.*, **1**, 69–76.

14

Protective measurement and the nature of the wave function within the primitive ontology approach

VINCENT LAM

14.1 Introduction

One of the crucial issues about protective measurements is the extent to which they provide (additional) grounds for some realist understanding of the wave function. This issue is notoriously subject to debate, which is not the aim of this chapter. Rather, this chapter aims at clarifying the further issue of the ontological picture corresponding to the realist understanding of the wave function possibly suggested by protective measurements. Oddly enough, proponents of a realist reading of protective measurements remain in general rather vague about ontology (see for instance Aharonov et al. (1993, 4624), who assimilate the wave function to an "extended object" without further details, or the brief and somewhat unprecise discussion in Dickson (1995, 135–136) of the interpretation of the wave function within Bohmian mechanics (BM) and within the theoretical framework elaborated by Ghirardi, Rimini and Weber (GRW); in this volume, Gao does propose a more elaborated ontological model for quantum mechanics (QM) based on protective measurements).

Despite this vagueness (or maybe because of it), it seems that the most obvious ontological intuition resulting from protective measurements points towards some form of straightforward realism about the wave function understood in the sense of a real, physical field: indeed, protective measurements are claimed to allow measuring expectation values of observables on a single quantum system, thereby providing the possibility of reconstructing (i.e. "measuring" in some sense) the wave function of a single quantum system. Aharonov et al. (1996, 125) clearly express this intuition: "We can observe the expectations values of operators, and we can observe the density and the current of the Schrödinger wave. We can "see" in some sense the Schrödinger wave. This leads us to believe that it has physical reality."

Therefore, this kind of intuition seems in line with a realist conception of the wave function, according to which the wave function is considered as a real (substantial, material in some sense) entity on its own (this view is often called "wave function realism"). Despite its natural appeal (that is, realist attitude with respect to the central theoretical entity of an empirically very successful theory – moreover, according to the proponents of protective measurement, the theoretical entity in question is not a mere statistical device, but has direct observable significance), such a realist understanding of the wave function faces several difficulties. Of course, the main one is that the wave function is not defined on ordinary 3-dimensional space but on a higher dimensional space (the dimension of which depends on the number of particles involved), so that considering the wave function as a real, physical field (often) implies admitting this so-called "configuration space" in the fundamental ontology of QM. Although there is nothing incoherent with such an abstract ontology itself, the main challenge is to provide an account in this context of our daily "illusion" of macroscopic objects in 3-dimensional space and evolving in time.

By contrast to the abstract ontology offered by wave function realism, there is a recently much discussed approach to the ontology of QM according to which the theory – and possibly any fundamental physical theory – is ultimately about entities in 3-dimensional space (or 4-dimensional spacetime) and their temporal evolution. Such an ontology postulating from the start matter localized in "usual" physical space or spacetime is called "primitive ontology" in the recent literature on the topic. According to the proponents of the primitive ontology approach to QM, it is the best (realist) way to avoid the main difficulty of wave function realism: there is no "illusion" or "appearance" of matter in 3-dimensional space to be explained, since this fact is simply postulated from the start as the referent of the theory (i.e. as what the theory is fundamentally about). Of course, this "postulate" is not the whole story: the theory in question has to specify how matter is instantiated in 3-dimensional space (particles, fields, strings, loops, ...) and how it evolves in time. And that's where things get interesting: in all the proposed primitive ontologies for QM (the paradigmatic examples are of course BM and versions of GRW), the wave function plays a central and crucial role in the time evolution of the primitive ontology and in the account of non-locality. An important and difficult task for the primitive ontologist is therefore to elucidate what the status of the wave function can be in this context.

The aim of this contribution is to discuss the nature of the wave function and in particular features revealed by protective measurements within the framework of the primitive ontology approach to QM. The primary aim is not to argue in favor of (or against) the primitive ontology approach to QM (see Ney and Albert (2013) for a recent review of the debates around primitive ontology and wave function realism); rather, the idea is to discuss protective measurements and the status of the wave function in an ontologically serious way, and the primitive ontology approach offers an interesting opportunity to do so. We discuss in Section 14.2 the motivation for the primitive ontology approach to QM and how a certain tension arises in this context about the status of the wave function. In Section 14.3 we suggest that an interpretation of the wave function within the framework of ontic structural realism, providing a clear and coherent ontological picture, resolves this tension and sheds some interesting light on the meaning of protective measurements and related experimental set-up. A conclusion and a few broader comments follow in Section 14.4.

14.2 Primitive ontology and the nature of the wave function

14.2.1 The main motivation for a primitive ontology

A primitive ontology for QM specifies explicitly what the theory is about, i.e. what there is in the world according to QM, in terms of material entities localized in 3-dimensional space (or 4-dimensional spacetime) and their dynamics. There are different such specifications, in particular within two of the three standard realist conceptions of QM that take the measurement problem seriously: particles following continuous deterministic trajectories within the framework of BM, a continuous stochastic mass density field or stochastic point-like events ("flashes") within GRW (giving rise to two versions of GRW: GRWm and GRWf). A primitive ontology in the context of the Everett (or "many-worlds") framework can possibly be defined (for instance in terms of a deterministic mass density field, see Allori et al., 2011) but its meaning is less transparent.

The main motivation for specifying a primitive ontology for QM is rather straightforward: it provides a powerful and generic explanatory framework within which familiar macroscopic objects localized in 3-dimensional space and their (classical) behavior can be understood in terms of the behavior of (possibly fundamental) microscopic entities that are also localized in 3-dimensional space (in particular, there is an explicit connection between the behavior of these microscopic entities and what can be observed at the macroscopic level, for instance in terms of measurement outcomes). Obviously, the details of this account depend on the specific primitive ontology under consideration; the point is that such an account in terms of a primitive ontology does not have to bridge substantial explanatory gaps, for instance between macroscopic objects that are (or seem to be) localized in 3-dimensional space and fundamental microscopic entities that are not.

It is interesting to note that the primitive ontology approach to QM finds part of its roots in Bell's notion of "local beables", which was introduced in the context of his reflections on non-locality and the measurement problem (see the papers collected in Bell, 1987). To some extent, a primitive ontology is made up of local beables – that is, of entities that "can be assigned to some bounded spacetime region" (Bell, 1987, 53) – which can be directly related to the behavior of familiar material objects, such as a measurement apparatus for instance, so that the measurement problem simply does not arise within the framework of a primitive ontology for QM (in this sense, the main motivation for a primitive ontology for QM is that there is simply no quantum measurement problem in this context).

14.2.2 *The central role of the wave function*

However, the local beables alone (in 3-dimensional physical space) have no explanatory power; the material entities that are localized in 3-dimensional physical space and that constitute the primitive ontology have no explanatory power alone. The explanatory power of the primitive ontology approach to QM stems from the local beables together with their temporal development or dynamics, which crucially relies on the wave function. As a consequence, in this context, it seems unavoidable to accept the wave function on top of or as "part of" the primitive ontology, in some sense to be clarified (note that in principle it does seem possible to consider the primitive ontology move within a purely Humean framework where all quantum features, including the wave function and the features related to quantum non-locality, supervene on the entire spacetime distribution of local beables, see Miller (2013) – however, the standard difficulties related to the explanatory power of Humeanism seem especially salient in the quantum case).

Let us consider BM as an illustration. Indeed, BM embodies the paradigmatic example of a primitive ontology for QM; it will serve us as a very convenient study case throughout this chapter. The Bohmian particles constitute the primitive ontology (they obviously are local beables since they always have a definite position in 3-dimensional physical space), but the temporal evolution of the total configuration of the Bohmian particles crucially relies on the universal wave function through the Bohmian guiding equation or equation of motion. According to this equation, Bohmian particles continuously evolve along determinate trajectories, but such that the velocity of each particle depends on the positions of all the other particles: strictly speaking the velocity of each particle is a functional of the universal wave function defined on the whole configuration. In particular, the role of the wave function in this huge dynamical interdependence is central to the Bohmian account of quantum non-locality, that is, to its explanatory power (more on that below). Clearly, this crucial role of the wave function is shared by the other main primitive ontologies of QM (e.g. GRWm and GRWf); the wave function is an irreducible part of what Allori et al. (2008) have identified as the "common structure" of all the conceptions within the primitive ontology approach to QM.

The "common structure" between GRWm, GRWf and BM is that the considered theory is fundamentally about matter in spacetime, in contrast to, e.g., a wave function in a high-dimensional configuration space; however, the point here, which is rather clear among the proponents of a primitive ontology for QM, is that the wave function cannot be entirely dropped from the ontological picture (see, however, the investigations in Dowker and Herbauts, 2005). The theoretical, explanatory, ontological importance of the wave function therefore creates a tension within the primitive ontology approach to QM: it raises the (old) issue of the ontological status and the metaphysical nature of the wave function within the familiar ontological picture offered by the primitive ontology framework, that of matter localized in 3-dimensional space and evolving in time.

14.2.3 Primary and secondary ontology

Before addressing the options available for the status of the wave function within the primitive ontology approach to QM, it is interesting to mention a possible strategy that has been put forward in order to alleviate this tension. The idea is to make a distinction between a primary part and a secondary part of the ontology of QM (see Maudlin, 2013; see also Allori, 2013): the primary part is the part of the ontology that is "more directly and unproblematically related to empirical data", whereas the secondary part is "more remote from empirical data, and hence more speculative" (Maudlin, 2013, 144). Within BM, the primary part of the ontology is constituted by the Bohmian particles, which make up the measurement apparatuses displaying definite measurement outcomes and, more generally, which ultimately make up all empirical data; in this sense, the wave function does not belong to the primary ontology, and its nature is therefore "more speculative". It is interesting to note that the first reason Maudlin (2013, 148) mentions for not considering the wave function (or quantum state) as part of the primary ontology is that "no experiment can *reveal* or *determine* the exact quantum state of a given system". Prima facie, protective measurements seem to provide a counterexample to this last claim, since a "sufficient number of protective measurements performed on a single system allow measuring its quantum wave function" (Vaidman, 2009, 506). So, on the one hand, protective measurements seem to speak in favor of considering the wave function (or quantum state) as part of the primary ontology of QM; but, on the other hand, even within the framework of protective measurements, the epistemic access to the wave function is indirect in the sense that it is mediated by the primary ontology (e.g. Bohmian particles) making up the empirical data (moreover, strictly speaking, protective measurements do not determine the *exact* wave function or quantum state, but only up to a phase). In any case, the distinction between the primary and secondary parts of the ontology of QM is epistemic, not ontological: in one way or another, in the context of the primitive ontology approach to QM, all agree that the wave function is part of the ontology of the theory, so that the fundamental issue of its metaphysical nature cannot be avoided.

14.2.4 The nature of the wave function

There are mainly three realist ways to understand the wave function as part of the ontology of QM (see Belot, 2012), two of which are common within the primitive ontology approach. The first one is the most straightforward: to consider the wave function as a physical object on its own. It is also the most problematic from the point of view of the primitive ontology approach. Indeed, within the framework of BM, it amounts to recognizing the wave function as a physical object, possibly not living in 3-dimensional space but in high-dimensional configuration space, in addition to the Bohmian particles in 3-dimensional space, thereby considerably inflating the ontology. The explanatory strength and simplicity of the primitive ontology approach would then be significantly weakened, and some of the difficulties of wave function realism – against which the primitive ontology approach was originally designed – would reappear. It is therefore not surprising that this option is commonly rejected within the framework of the primitive ontology approach to QM.

The second understanding rather suggests to consider the wave function as a law-like, nomological entity, that is, not as an additional substantial, physical entity in space and time. This interpretation of the wave function is favored by some of the most prominent current proponents of BM, who take as a heuristic argument the analogy with the common interpretation of the Hamiltonian on phase space within the framework of classical mechanics (see e.g. Dürr et al., 1997). So, within this understanding and the Bohmian context, the wave function is taken as an aspect of the Bohmian law of motion (guiding equation). However, this nomological interpretation of the wave function faces an important difficulty: the wave function can be time-dependent – a non-standard feature for a law-like entity, which requires some clarifications. A related difficulty concerns the status of the Schrödinger equation: what is the status of a law (the Schrödinger equation) that determines the temporal evolution of a law-like entity (the wave function)? In order to deal with these difficulties, the proponents of the nomological understanding of the wave function within BM have deployed a strategy which contains three main components. First, the crucial distinction between the universal wave function, i.e. the wave function of the Universe (the wave function corresponding to all the Bohmian particles in the Universe), and the effective wave functions corresponding to (Bohmian) subsystems of the Universe. If the latter are epistemically crucial (they are the ones that are dealt with in standard QM as well as for predictive and operational purposes), only the former is ontologically fundamental strictly speaking (effective wave functions only possess a "derivative" status compared to the universal wave function). Second, the expectation, based on the fundamental timelessness of the Wheeler–DeWitt equation in canonical quantum gravity, that the universal wave function is static (and possibly unique). Third, the (informed) conjecture that the time-dependent Schrödinger evolution is not fundamental, but only effective in the sense of only arising for subsystems and their effective ("time-dependent") wave functions. Even if supported by good heuristic arguments, the second and third components of this strategy remain speculative, making the purely nomological interpretation of the wave function somewhat less attractive. Moreover, it seems that the exact ontological picture resulting from the nomological understanding of the wave function depends on one's metaphysical stance with respect to laws, e.g. Humean or dispositional. There is no need to enter this venerable metaphysical debate here. Suffice it to note that if a Humean approach in this context is clearly not incoherent (leading to what could be called "quantum Humeanism"), it can be argued that it would considerably weaken the explanatory power of the conception, which is one of the main motivations for the primitive ontology move in the first place. For instance, within this Humean framework, the wave function and crucial quantum features such as quantum non-locality that are encoded in the wave function would merely supervene on the whole distribution of the relevant local beables in the entire spacetime, rather than being anchored (and therefore "explained" in some sense) in the nature or properties of these local beables postulated by the primitive ontology (see the comment above in Section 14.2.2; see also the discussion in Esfeld et al., 2013).

The third understanding of the wave function precisely aims to do that: the idea is to interpret the wave function in terms of the properties of – more precisely, the relations among – the local beables. This understanding is appealing in the primitive ontology context: indeed, for example, within the framework of BM, the wave function determines through the Bohmian equation of motion (guiding equation) the temporal development of the local beables, that is, the velocities of the Bohmian particles. In this perspective, it is perfectly sensible to think of the wave function as describing a fundamental property of the Bohmian particles that determines their motion (this description possibly not being one-to-one, see Belot, 2012, 78–80 and the discussion in Esfeld et al., 2013, Section 4). The main worry for this understanding comes from the fact that, in this context, the wave function encodes quantum non-locality, so that the fundamental property described by the wave function is rather peculiar. In the Bohmian case, the (universal) wave function is defined on the whole configuration of all Bohmian particles in the Universe at a given time, so that the temporal development (the velocity) of each particle depends strictly speaking on the positions of all the other particles at that time through the (universal) wave function. Therefore, the wave function actually describes a kind of holistic property of the whole configuration of particles (at a given time). We discuss in the next section how ontic structural realism can help to clarify the nature of such a property, that is, the nature of the wave function in the primitive ontology approach.

14.3 Quantum structure

14.3.1 Ontic structural realism and primitive ontology

Ontic structural realism (OSR) is a recently much debated conception in the metaphysics of contemporary fundamental physics, in particular quantum theory. As a metaphysical conception and interpretative framework for fundamental physics, its development has been mainly motivated by various fundamental relational physical features, in particular background independence and gauge-theoretic diffeomorphism invariance in the general relativistic domain (see e.g. the contributions by Pooley, Rickles and Stachel in Rickles et al., 2006, as well as Esfeld and Lam, 2008) and permutation invariance (together with other symmetry considerations), entanglement and non-locality in the quantum domain (see e.g. French and Ladyman, 2003, Esfeld, 2004, Ladyman et al., 2007, chapter 3, Kantorovich, 2009, Muller, 2011 and Lam, 2013). The broad ontological thesis of OSR that is motivated by these relational features can be expressed in the following way: what there is in the world at the fundamental level (or in the cases where OSR is relevant) are physical structures, in the sense of networks of concrete physical relations among concrete physical objects (relata), whose existence depends in some sense on relations in which they stand (on structures they are part of).

As mentioned above, it has been argued in the literature for some time now that OSR provides a general interpretative framework for the generic relational features of quantum entanglement and quantum non-locality as encoded in the violations of Bell-type inequalities. Since (of course) quantum non-locality has to be accounted for within the primitive ontology approach, it is no wonder that OSR is relevant in this context. Indeed, on the one hand, a primitive ontology for QM (such as BM) provides an ontological framework within which the general OSR understanding of quantum non-locality can be specified (in particular the relata of the relevant quantum structures can be specified). On the other hand, OSR provides the primitive ontology approach to QM with a convincing way to interpret the wave function and its encoding of quantum non-locality in this context (see Esfeld, 2014, for a similar point of view).

14.3.2 The wave function as a physical structure

We have seen above that there is a tension about the wave function within the primitive ontology approach to QM: if it clearly plays a central role in the explanatory scheme of the primitive ontology approach (in particular, in the account of non-locality), it is not easy to anchor the wave function in the 3-dimensional physical space (or 4-dimensional spacetime) in which the relevant local beables live – as one would expect within the primitive ontology framework. OSR precisely provides such a spacetime anchorage for the wave function in the primitive ontology approach to QM: in this context, the wave function can be understood as a physical structure in spacetime whose relata are the local beables of the primitive ontology under consideration (this spacetime anchorage is crucial from the primitive ontology point of view; for a more abstract and controversial OSR interpretation of the wave function, see French, 2013).

Let us illustrate how this account works in our study case, BM. The wave function is understood in terms of physical relations among all the Bohmian particles. This huge network of physical relations (i.e. the physical structure described by the wave function) constitutes the physical ground for (the explicit) quantum non-locality in BM; in this fundamental quantum structure, each particle is strictly speaking related to all the others (in a way described by the wave function), so that its temporal development (its velocity) depends on the positions of all the other particles (hence the BM account of the violation of Bell-type inequalities in terms of the violation of parameter independence); note that, as mentioned in Section 14.2.4, the notion of effective wave function captures the operationally relevant aspects of such a huge dependence.

There are two aspects that jointly make this quantum structure, which is described in the quantum formalism by the wave function, a structure in the OSR sense. First, the quantum relations connecting all the Bohmian particles do not supervene on any intrinsic properties of the particles; therefore, these quantum relations and the corresponding quantum structure are fundamental and irreducible in the sense that they cannot be merely understood in terms of (they cannot be "reduced" to) the intrinsic properties of the relata, namely the Bohmian particles. Second, even if some intrinsic individuality and identity can possibly be ascribed to them (e.g. in virtue of their spacetime location, if one accepts that it can be taken as an intrinsic feature), there is a sense in which Bohmian particles dynamically depend on the structure they are part of, through the dependence on the positions of all the other particles. Unlike the case of Newtonian gravity (where some structuralist dependence among all Newtonian particles also obtains), this dependence is strictly speaking not affected by spatial distance. So, in a sense, the very existence of Bohmian particles dynamically depends on the structure they are part of.

Furthermore, one could characterize Bohmian particles in terms of some dynamical (diachronic) identity that depends on the whole configuration of particles, that is, in terms of some non-intrinsic (structural, contextual) identity (about the notion of non-intrinsic identity in the context of OSR, see Lam, 2014). The tension between this structural identity and the above mentioned intrinsic identity based on the spacetime location is only apparent: besides the fact that its "intrinsicness" can be put into question (it ultimately relies on the spacetime structure), this latter identity is dynamically inert, whereas the former plays a crucial dynamical and explanatory role, in particular in the account of quantum non-locality. Indeed, in this context, quantum non-locality is accounted for in terms of the dynamics of the (relevant part of the) quantum structure, within which the relata (i.e. the Bohmian particles) are interdependent; this dynamical interdependence is here precisely encoded in the notion of dynamical (or structural, contextual) identity. In this perspective, it seems to make sense to claim that for each Bohmian particle, the fact of being this very particle, which includes its own trajectory and dynamical features, depends on the structure it is part of.

From the metaphysical point of view, the structural account of the wave function proposed here provides a clear metaphysical basis for the holistic aspects mentioned within the framework of the property understanding at the end of Section 14.2.4; in this sense, OSR helps to clarify the nature and the status of the wave function – more precisely: what is represented by the wave function – within the primitive ontology approach to QM. We now consider protective measurements in this structuralist and primitive ontology (in particular, BM) context; more specifically, we discuss to what extent this structuralist point of view might help to clarify the alleged difficulties for BM posed by protective measurements.

14.3.3 Protective measurements and primitive ontology: probing the quantum structure

As mentioned in Section 14.1, protective measurements are often (somewhat vaguely) understood as providing grounds for a realist interpretation of the wave function, in the sense of directly probing the wave function itself (on a single system). However, from a metaphysical point of view, protective measurements by themselves cannot illuminate the nature of the wave function as long as the measurement problem is not addressed (note that, despite their strong claims about the reality of the wave function, Aharonov et al., 1996, 121, are clear that protective measurements do not help to solve the measurement problem). But in the cases where a clear ontology for QM is provided (that is, in the cases where the measurement problem is addressed, as within the primitive ontology framework), protective measurements provide an interesting tool to investigate the nature of the wave function. As Vaidman (2009, 506) puts it: "The protective measurement procedure is not a proof that we should adopt one interpretation instead of the other, but it is a good testbed which shows advantages and disadvantages of various interpretations." In this perspective, it is natural to consider protective measurement in the context of the primitive ontology approach to QM.

Indeed, protective measurements and weak measurements in general (together with the "which-way" experiment set-up) have been extensively discussed in the context of BM; in particular, they are supposed to highlight some unsatisfactory features of BM (see e.g. Englert et al., 1992, Aharonov and Vaidman, 1996, Aharonov et al., 1999). For instance, an important part of these thought experiments crucially aims to show that an empty wave of a particle (i.e. a part of the wave function without the actual particle position in its domain, e.g. after crossing a beam splitter) can have observable effects, such as a trace of bubbles in a (slowly developing) bubble chamber (leading to what has been called a "surrealistic trajectory" in Englert et al., 1992; see also Vaidman, 2005 and references therein) or a straightforward modification of the pointer's position of a measuring device. These alleged difficulties all ultimately rely on the above-mentioned tension about the status of the wave function in the primitive ontology approach to QM, in particular BM (see the end of Section 14.2.2). Indeed, the fathers of protective measurements are themselves fully aware of this fact when they write: "Note, however, that the difficulties we see follow mostly from a particular approach to the Bohm theory we adopt, in which only the Bohmian particles correspond to the "reality" which we experience, while the wave function is just a pilot wave which governs the motion of the particle" (Aharonov and Vaidman, 1996, 141).

The OSR understanding of the wave function within the primitive ontology (in particular, BM) framework elaborated in the preceding section provides the wave function with a clear and robust ontological status that dissolves the above-mentioned tension and that unproblematically accounts for weak and protective measurements (as well as the related which-way measurement set-up). From this structuralist point of view, these thought experiments can be merely understood as specific ways to "directly" probe the quantum structure (in the precise OSR sense) represented by the wave function (to some extent, within this framework, any quantum measurement is a probing of the quantum structure through the behavior of its relata, the relevant local beables). It is important to see that this direct probing is unproblematic in the OSR and primitive ontology context, since the quantum structure represented by the wave function is a concrete physical structure in the usual 3-dimensional space among the local beables under consideration (e.g. Bohmian particles); in particular the worries related to the possibility of directly observing an abstract wave function living in $3N$-dimensional space simply do not arise here (for such worries in the Bohmian case, see for instance Boscá, 2013, 56).

These thought experiments highlight in particular non-local effects in BM that are at the roots of what the critics consider to be mysterious ("surrealistic") and unsatisfactory features. Let's see how this comes about and how the OSR understanding of the wave function might dispel these worries. Very roughly, the non-local features displayed in these measurement procedures stem from the fact that, in these rather specific cases, the wave function corresponding to a Bohmian particle can have an observable (measurable) effect where the Bohmian particle is *not* located (note that these non-local features have to be distinguished from the ones arising in the context of the violation of Bell-type inequalities, although they all emanate from the wave function in the end). At first sight, this fact seems to be very surprising, especially for "position measurements" recording what looks like a trajectory (e.g. a trace in a bubble chamber) that completely differs from the actual trajectory of the particle in question; hence the doubts about the reality and meaning of the actual Bohmian positions and trajectory. Similarly, the protective measurement set-up (roughly, measurement as a very weak "adiabatic" interaction during a very long time) for the position of a Bohmian particle at rest shows that the position measuring device can be triggered (i.e. the pointer's position is modified) in regions where the Bohmian particle is *not* located (so that the device seems to record the position of the particle where it is not – if this makes any sense).

Of course, the last part of the last sentence does not make much sense. Three important and interrelated points help to clarify the situation and highlight in particular the role of the structuralist interpretation of the wave function in this clarification. First, the notion of position measurement in the Bohmian context is a subtle issue and we should be careful about it; for instance, Drezet (2006) argues that we are not really dealing with a genuine Bohmian position measurement in the above thought experiment involving a protective measurement (there is no local interaction with the Bohmian particle), but rather a measure of some quantity (density) directly related to the wave function itself (on this issue in the context of "surrealistic trajectories", see also Barrett, 2000). Second, as already mentioned above, these at first sight surprising effects are best understood as non-local features of the relevant wave function (e.g. the wave function corresponding to the particle and the pointer); it is interesting to note that most of the commentators agree on this understanding in terms of the wave function, whether or not they find it satisfactory (see e.g. Dewdney et al., 1993, Barrett, 2000, Vaidman, 2005, Drezet, 2006, and Boscá, 2013). Third, and most importantly for our aim here, the structuralist interpretation of the wave function as a concrete physical structure (in the precise OSR sense) does provide an entirely satisfactory account of the non-local features displayed in these thought experiments. Indeed, from this structuralist point of view, these non-local features are due to the inherently relational nature of the quantum structure (represented by the wave function) under consideration (see Section 14.3.2). For instance, in the protective measurement case, the Bohmian particle that is being measured and the Bohmian particles constituting the pointer of the measuring device are relata of the same quantum structure, which is represented by the relevant wave function. The fact that, from the ontological point of view, one should, strictly speaking, only consider the universal wave function (see Section 14.2.4) does not alter the main point here – indeed, Norsen and Struyve (2013) have recently convincingly suggested that what is being measured in weak measurement procedures involving entangled systems precisely corresponds to effective (conditional) wave functions. Now, the temporal development (of the relevant quantum structure) corresponding to the protective measurement interaction (in the region where the measured Bohmian particle is not located) is such that the position of the pointer is modified through this interaction process and even though the region in question does not contain the Bohmian particle that is being measured (see the Bohmian analysis of the "wave function induced" motion of the particle and the pointer in Drezet, 2006). Obviously, a similar structuralist account can be provided in the case of "surrealistic" Bohmian trajectories.

The physics of the Bohmian account of these thought experiments is clear and coherent – as pointed out in the early reaction of Dürr et al. (1993) to Englert et al. (1992), it is interesting to note that no one actually claims that BM is shown to be incoherent in these thought experiments, the objections being rather at the metaphysical level. The structuralist understanding of the wave function suggested here within the Bohmian account provides a clear and coherent ontological picture of what is going on in these thought experiments – a convincing ontological picture that dispels the metaphysical worries: these thought experiments are specific ways to directly probe the relevant concrete quantum structure.

14.4 Conclusion and perspectives

In this chapter, we have discussed the nature of the wave function (for instance, as unveiled by protective measurements) within the framework of the primitive approach to QM, that is, within the realist interpretations of the theory – realist solutions to the measurement problem – according to which QM is ultimately about (material) entities localized in 3-dimensional physical space and evolving in time. Within this primitive ontology framework, we have suggested a structuralist understanding of the wave function in the sense of OSR and we have considered protective measurements in this novel interpretative light. We have mainly taken BM – the paradigmatic example of a primitive ontology for QM – as a convenient study case for illustrating the interpretative relevance of this structuralist understanding. In particular, the explicit non-locality that is at the heart of the seemingly counter-intuitive features revealed by protective measurements in the BM context is naturally understood in terms of the relational features of the relevant quantum structure. On this basis, we have suggested a clear interpretation of protective measurements (and related weak measurements and other experimental set-ups involving "surrealistic trajectories") in the framework of BM according to which these specific measurement procedures are best understood as specific ("direct") ways to probe the quantum structure (in a precise sense) represented by the relevant (effective, conditional) wave function.

This interpretation of the wave function as representing a concrete physical structure instantiated in spacetime can clearly be specified within the other standard primitive ontologies that have been proposed for QM, such as GRWm and GRWf in particular. Indeed, the wave function is an irreducible part of what Allori et al. (2008) have identified as the "common structure" of GRWm, GRWf and BM (see Section 14.2.2). It would actually be an interesting project to specify in detail the quantum structure represented by the wave function within the framework of the other possible primitive ontologies besides BM in order to highlight the differences among these realist interpretations of QM (in their treatment of quantum non-locality in particular).

Despite the intuitive appeal (e.g. in terms of explanatory power) of the primitive ontology approach to QM, we would like to conclude on a cautionary note. As discussed in this chapter, such an ontology postulates from the start matter localized in 3-dimensional space and evolving in time (or localized in 4-dimensional spacetime); and, as discussed in this chapter, there are good reasons for doing so in the context of QM, which is the primary target of this ontological and interpretative move – reasons very similar to Bell's motivation for introducing local beables. At this point, all is fine. However, there are some expectations that the primitive ontology framework remains valid for any fundamental theory, and in particular any fundamental quantum theory. Now, when quantum theory is applied to the gravitational field as described by the general theory of relativity (of course, we still don't have a complete quantum theory of the gravitational field, but only various research programs; note that there is no compelling reasons for quantizing the gravitational field, but only strong expectations), there is an on-going debate in both physics and philosophy communities about the status of spacetime itself, which is partly due to the very nature of the gravitational field within the general theory of relativity. Without entering into this debate, we suggest that postulating matter localized in spacetime from the start when dealing with the interpretation (and the ontology) of a theoretical framework (that of quantum gravity) where the very status of spacetime and its relationship to matter constitute crucial open issues does not seem to be the right methodology.

Acknowledgements

I would like to thank Shan Gao for the invitation to contribute to this volume and for his patience. I am grateful to the Swiss National Science Foundation (Ambizione grant PZ00P1_142536/1) for financial support.

References

Y. Aharonov, J. Anandan, and L. Vaidman. Meaning of the wave function. *Physical Review A*, **47**(6):4616–4626, 1993.

Y. Aharonov, J. Anandan, and L. Vaidman. The meaning of protective measurements. *Foundations of Physics*, **26**(1):117–126, 1996.

Y. Aharonov, B.-G. Englert, and M. O. Scully. Protective measurements and Bohm trajectories. *Physics Letters A*, **262**:137–146, 1999.

Y. Aharonov and L. Vaidman. About position measurements which do not show the Bohmian particle position. In J. T. Cushing, A. Fine, and S. Goldstein (eds.), *Bohmian Mechanics and Quantum Theory: an Appraisal*, pages 141–154. Dordrecht, Springer, 1996.

V. Allori. Primitive ontology and the structure of fundamental physical theories. In A. Ney and D. Z. Albert (eds.), *The Wave Function: Essays in the Metaphysics of Quantum Mechanics*, pages 58–75. New York, Oxford University Press, 2013.

V. Allori, S. Goldstein, R. Tumulka, and N. Zanghì. On the common structure of Bohmian mechanics and the Ghirardi–Rimini–Weber theory. *British Journal for the Philosophy of Science*, **59**:353–389, 2008.

V. Allori, S. Goldstein, R. Tumulka, and N. Zanghì. Many worlds and Schrödinger's first quantum theory. *British Journal for the Philosophy of Science*, **62**:1–27, 2011.

J. A. Barrett. The persistence of memory: surreal trajectories in Bohm's theory. *Philosophy of Science*, **67**:680–703, 2000.

J. S. Bell. *Speakable and Unspeakable in Quantum Mechanics*. Cambridge, Cambridge University Press, 1987.

G. Belot. Quantum states for primitive ontologists. *European Journal for Philosophy of Science*, **2**:67–83, 2012.

M. C. Boscá. Some observations upon "realistic" trajectories in Bohmian quantum mechanics. *Theoria*, **76**:45–60, 2013.

C. Dewdney, L. Hardy, and E. J. Squires. How late measurements of quantum trajectories can fool a detector. *Physics Letters A*, **184**:6–11, 1993.

M. Dickson. An empirical reply to empiricism: protective measurement opens the door for quantum realism. *Philosophy of Science*, **62**(1):122–140, 1995.

F. Dowker and I. Herbauts. The status of the wave function in dynamical collapse models. *Foundations of Physics Letters*, **18**:499–518, 2005.

A. Drezet. Comment on "Protective measurements and Bohm trajectories". *Physics Letters A*, **350**:416–418, 2006.

D. Dürr, S. Fusseder, S. Goldstein, and N. Zanghì. Comment on "Surrealistic Bohm trajectories". *Zeitschrift für Naturforschung*, **48a**:1261–1262, 1993.

D. Dürr, S. Goldstein, and N. Zanghì. Bohmian mechanics and the meaning of the wave function. In R. Cohen, M. Horn, and J. Stachel (eds.), *Experimental Metaphysics*, pages 25–38. Dordrecht, Kluwer, 1997.

B.-G. Englert, M. O. Scully, G. Süssmann, and H. Walther. Surrealistic Bohm trajectories. *Zeitschrift für Naturforschung*, **47a**:1175–1186, 1992.

M. Esfeld. Quantum entanglement and a metaphysics of relations. *Studies in History and Philosophy of Modern Physics*, **35**:601–617, 2004.

M. Esfeld. How to account for quantum non-locality: ontic structural realism and the primitive ontology of quantum physics. *Synthese*, to appear, 2014.

M. Esfeld and V. Lam. Moderate structural realism about space-time. *Synthese*, **160**:27–46, 2008.

M. Esfeld, D. Lazarovici, M. Hubert, and D. Dürr. The ontology of Bohmian mechanics. *British Journal for the Philosophy of Science*, **64**:doi:10.1093/bjps/axt019, 2013.

S. French. Whither wave function realism. In A. Ney and D. Z. Albert (eds.), *The Wave Function: Essays in the Metaphysics of Quantum Mechanics*, pages 76–90. New York, Oxford University Press, 2013.

S. French and J. Ladyman. Remodelling structural realism: quantum physics and the metaphysics of structure. *Synthese*, **136**(1):31–56, 2003.

A. Kantorovich. Ontic structuralism and the symmetries of particles physics. *Journal for General Philosophy of Science*, **40**:73–84, 2009.

J. Ladyman, D. Ross, D. Spurett, and J. Collier. *Every Thing Must Go: Metaphysics Naturalized*. Oxford, Oxford University Press, 2007.

V. Lam. The entanglement structure of quantum field systems. *International Studies in the Philosophy of Science*, **27**:59–72, 2013.

V. Lam. Entities without intrinsic physical identity. *Erkenntnis*, doi:10.1007/s10670–014–9601–5, 2014.

T. Maudlin. The nature of the quantum state. In A. Ney and D. Albert (eds.), *The Wave Function*, Oxford, Oxford University Press, 2013.

E. Miller. Quantum entanglement, Bohmian mechanics, and Humean supervenience. *Australasian Journal of Philosophy*, doi:10.1080/00048402.2013.832786, 2013.

F. A. Muller. Withering away, weakly. *Synthese*, **180**:223–233, 2011.

A. Ney and D. Z. Albert (eds.). *The Wave Function: Essays in the Metaphysics of Quantum Mechanics*. New York, Oxford University Press, 2013.

T. Norsen and W. Struyve. Weak measurements and (Bohmian) conditional wave functions. *arXiv:1305.2409v2*, 2013.

D. Rickles, S. French, and J. Saatsi. *The Structural Foundations of Quantum Gravity*. Oxford, Oxford University Press, 2006.

L. Vaidman. The reality in Bohmian quantum mechanics or can you kill with an empty wave bullet? *Foundations of Physics*, **35**:299–312, 2005.

L. Vaidman. Protective measurements. In D. Greenberger, K. Hentschel, and F. Weinert (eds.), *Compendium of Quantum Physics*, pages 505–508. Berlin, Springer, 2009.

15

Reality and meaning of the wave function

SHAN GAO

The wavefunction gives not the density of stuff, *but gives rather (on squaring its modulus) the density of probability. Probability of what, exactly? Not of the electron* being *there, but of the electron being* found *there, if its position is 'measured'. Why this aversion to 'being' and insistence on 'finding'? The founding fathers were unable to form a clear picture of things on the remote atomic scale.*

John S. Bell (1990)

15.1 Introduction

The physical meaning of the wave function is an important interpretative problem of quantum mechanics. Notwithstanding nearly ninety years of development of the theory, it is still an unsolved issue. During recent years, more and more research has been done on the ontological status and meaning of the wave function (see, e.g. Monton, 2002; Lewis, 2004; Gao, 2011a, 2011b; Pusey, Barrett and Rudolph, 2012; Ney and Albert, 2013). In particular, Pusey, Barrett and Rudolph (2012) demonstrated that under certain non-trivial assumptions such as the preparation independence assumption, the wave function of a quantum system is a representation of the physical state of the system.[1] This poses a further question, namely whether the reality of the wave function can be argued without resorting to non-trivial assumptions. Moreover, a harder problem is to determine the ontological meaning of the wave function, which is still a hot topic of debate in the realistic alternatives to quantum mechanics such as the de Broglie–Bohm theory or Bohmian mechanics (Belot, 2012).

In this chapter, we will first give a clearer argument for the reality of the wave function in terms of protective measurements, which does not depend on non-trivial

---

[1] For more discussions about the Pusey–Barrett–Rudolph or PBR theorem, see Colbeck and Renner (2012); Lewis et al. (2012); Schlosshauer and Fine (2012, 2013); Leifer and Maroney (2013); Patra, Pironio and Massar (2013); Wallden (2013).

assumptions and can also overcome existing objections. Next, based on an analysis of the mass and charge properties of a quantum system, we will propose a new ontological interpretation of the wave function. According to this interpretation, the wave function of an N-body system represents the state of ergodic motion of N particles. Moreover, the ergodic motion of particles is discontinuous and random in nature, and the modulus squared of the wave function gives the probability density that the particles appear in certain positions in space.

15.2 On the reality of the wave function

The ontological status of the wave function in quantum mechanics is usually analyzed in the context of conventional impulsive measurements. Although the wave function of a quantum system is in general extended over space, one can only detect the system in a random position in space by an (impulsive) position measurement, and the probability of detecting the system in the position is given by the modulus squared of the wave function there. Thus it seems reasonable for a realist to assume that the wave function does not refer directly to the physical state of the system but only relates to the state of an ensemble of identically prepared systems. Although there are several interesting theorems such as the PBR theorem which reject this epistemic view of the wave function, these theorems always depend on some non-trivial assumptions. For example, the PBR theorem depends on a preparation independence assumption (Pusey, Barrett and Rudolph, 2012). By denying these non-trivial assumptions, one can still restore the epistemic view of the wave function. Moreover, it has been demonstrated that additional assumptions are always necessary to rule out the epistemic view of the wave function when considering only conventional impulsive measurements (Lewis et al., 2012).

Thanks to the important discoveries of Yakir Aharonov and Lev Vaidman et al., it has been known that there exist other kinds of quantum measurement such as weak measurements and protective measurements (Aharonov, Albert and Vaidman, 1988; Aharonov and Vaidman, 1990, 1993; Aharonov, Anandan and Vaidman, 1993). In particular, by a series of protective measurements on a single quantum system, one can detect the system in all regions where its wave function extends and further measure the whole wave function of the system (Aharonov and Vaidman, 1993; Aharonov, Anandan and Vaidman, 1993). During a protective measurement, the measured state is protected by an appropriate procedure (e.g. via the quantum Zeno effect) so that it neither changes nor becomes entangled with the state of the measuring device appreciably. In this way, such protective measurements can measure the expectation values of observables on a single quantum system, even if the system is initially not in an eigenstate of the measured observable, and the whole wave function of the system can also be measured as expectation values of certain observables.

Since the wave function of a single quantum system can be measured by a series of protective measurements, it seems natural to assume that the wave function refers directly to the physical state of the system. Several authors, including the discoverers of protective measurements, have given similar arguments supporting this implication of protective measurements for the ontological status of the wave function (Aharonov and Vaidman, 1993; Aharonov, Anandan and Vaidman, 1993; Anandan, 1993; Dickson, 1995; Gao, 2013a). However, these analyses have been neglected by most researchers, and they are also subject to some objections (Dass and Qureshi, 1999; Chapters 7 and 13 of this volume). Here we will first present a clearer argument for the reality of the wave function in terms of protective measurements, and then answer these objections.

According to quantum mechanics, we can prepare a single measured system whose associated wave function is $\psi(t)$ at a given instant t. The question is whether the wave function refers directly to the physical state of the system or merely to the state of an ensemble of identically prepared systems. As noted above, this question can hardly be answered by analyzing non-protective impulsive measurements of the system, by each of which one obtains one of the eigenvalues of the measured observable, and the expectation value of the observable as well as the value of $\psi(t)$ can only be obtained by calculating the statistical average of the eigenvalues for an ensemble of identically prepared systems. Now, by a protective measurement on the measured system, we can directly obtain the expectation value of the measured observable. Moreover, by a series of protective measurements of certain observables on *this* system, we can further obtain the value of $\psi(t)$. Since we can measure the wave function *only* from a single prepared system by protective measurements, the wave function represents the physical state of a single system. Similarly, the expectation values of observables are also properties of a single system.

There are two possible objections to the above conclusion that protective measurements support the reality of the wave function. The first is based on the requirement that the unknown state of a single system is measurable. It claims that since the unknown state of a single quantum system cannot be protectively measured, protective measurements do not have implications for the ontological status of the wave function (see, e.g. Unruh 1994). However, this requirement is too stringent (see also Chapter 10). If it were true, then no argument for the reality of the wave function including the PBR theorem could exist, because it is a well-known result of quantum mechanics that an unknown quantum state cannot be measured. On the other hand, it is also worth noting that protective measurements alone cannot imply the reality of the wave function (see also Chapters 7 and 13). In both the PBR theorem and the above argument, a realist view of the relationship between theory and reality is implicitly assumed, according to which the theoretical terms expressed in the language of mathematics connect to the entities existing in the physical world. Under this assumption, when preparing a physical system with a given associated wave function, the wave function refers either to the physical state of the system or to the state of an ensemble of identically prepared systems. As argued above, it is here that protective measurements help determine which interpretation is true. This determination does not require that the wave function of the prepared system is unknown beforehand.[2]

The second objection concerns realistic protective measurements (Dass and Qureshi, 1999; Chapter 13 of this volume). A realistic protective measurement can never be performed on a single quantum system with absolute certainty. For example, for a realistic protective measurement of an observable A on a non-degenerate energy eigenstate whose measurement interval T is finite, there is always a tiny probability proportional to $1/T^2$ of obtaining a different result $\langle A \rangle_\perp$, where $\perp$ refers to a normalized state in the subspace normal to the measured state as picked out by the first-order perturbation theory. It thus claims that the uncertainty precludes an ontological status for the wave function. If in the argument one directly resorts to the Einstein–Podolsky–Rosen criterion of reality (see, e.g. Chapter 10), according to which *"If, without in any way disturbing a system, we can predict with certainty (i.e. with probability equal to unity) the value of a physical quantity, then there exists an element of physical reality corresponding to this physical quantity."* (italics in the original) (Einstein, Podolsky and Rosen, 1935), then this objection may be valid. However, one may avoid this objection by resorting to a somewhat different criterion of reality, which seems more reasonable and also appropriate for realistic protective measurements.

The new criterion of reality is that if, with an arbitrarily small disturbance on a system, we can predict with probability arbitrarily close to unity the value of a physical quantity, then there exists an element of physical reality corresponding to this physical quantity. Although a realistic protective measurement with finite measurement time T can never be performed on a single quantum system with absolute certainty, the uncertainty and the disturbance on the measured system can be made arbitrarily small when the measurement time T approaches infinity. Thus, according to this criterion of reality, realistic protective measurements also support the reality of the wave function. Note that in order to argue for the reality of the

---

[2] It is worth emphasizing that knowing the wave function beforehand is not a weak point in our argument either (which is often misunderstood by some authors). The reason is that the wave function is only a mathematical object associated with the physical system, and we need to determine its physical meaning, e.g. whether or not it represents the physical state of the prepared system. (In this sense, although the wave function is known, the physical state of the system is still unknown, and what a protective measurement measures is also an unknown physical state.) As we will see later, protective measurements may not only help answer this question, but also be helpful for investigating the nature of the physical state described by the wave function, for which the existing no-go theorems such as the PBR theorem cannot help much (see also Chapter 9).

wave function in terms of protective measurements, it is not necessary to directly measure the wave function of a single quantum system, and measuring the expectation value of an arbitrary observable on a single quantum system is enough. If the expectation values of observables are physical properties of a single quantum system, then the wave function, which can be reconstructed from the expectation values of a sufficient number of observables, will also represent the physical property or physical state of a single quantum system. This will avoid the scaling problem (see Chapter 13).

In addition, it can be argued that the uncertainty of realistic protective measurements does not prevent them from measuring the actual state of a single quantum system. When a realistic protective measurement obtains a result, the state of the whole system including the measured system and the measuring device collapses to the state corresponding to this result (according to standard quantum mechanics), in which the state of the measured system and the state of the measuring device are correlated. When the result of a realistic protective measurement is wrong, such as $\langle A \rangle_\perp$ with probability proportional to $1/T^2$, the measured state collapses to the state $\perp$. In this case, the result of the protective measurement does not reflect the original measured state, but reflects the resulting new state. By contrast, when a realistic protective measurement obtains the right result, namely the expectation value of the measured observable in the measured state, the resulting state of the measured system is still the original measured state, and the result of the protective measurement reflects the original measured state. This means that, unlike conventional impulsive measurements, the uncertainty of realistic protective measurements does not prevent them from measuring the actual state of a single quantum system, which turns out to be represented by its wave function, though it makes the probability of doing so smaller than one.

Interestingly, we can also give another argument for ψ-ontology in terms of protective measurements, which is similar to the argument used by the PBR theorem (Pusey, Barrett and Rudolph, 2012). For two arbitrary (protected) non-orthogonal states of a quantum system, select an observable whose expectation values in these two states are different. Then the overlap of the probability distributions of the results of protective measurements of the observable on these two states can be arbitrarily close to zero (e.g. when the measurement interval T approaches infinity). If there exists a non-zero probability p that these two non-orthogonal states correspond to the same physical state λ, then when assuming the same λ yields the same probability distribution of measurement results as the PBR theorem assumes, the overlap of the probability distributions of the results of protective measurements of the above observable on these two states will be not smaller than p. Since p is a determinate number, this leads to a contradiction. This argument, like the previous one, only considers a single quantum system, and thus avoids the preparation independence assumption used by the PBR theorem[3]. Note that the above protective measurements on the two *protected* non-orthogonal states are the same.

Finally, we note that there might also exist other components of the underlying physical state, which are not measureable by protective measurements and not described by the wave function, e.g. the positions of the particles in the de Broglie–Bohm theory. In this case, according to our argument, the wave function still represents the underlying physical state, though it is not a complete representation. Certainly, the wave function also plays an epistemic role by giving the probability distribution of measurement results according to the Born rule. However, this role will be secondary and determined by the complete quantum dynamics that describes the measurement process, e.g. the collapse dynamics in dynamical collapse theories.

15.3 Meaning of the wave function

If the wave function represents the physical state of a single quantum system, then what physical state does it represent? In this section, we will further investigate the ontological meaning of the wave function. We will first analyze one-body systems and then analyze many-body systems.

15.3.1 One-body systems

For a one-body quantum system, its spatial wave function in position x at instant t, $\psi(x, t)$, represents the physical state of the system in position x at instant t. This means that for a one-body system, there is a physical entity spreading out over a region of space where the spatial wave function of the system is not zero.[4] In the following, we analyze the existing form of the physical entity. The analysis may provide an important clue to the ontological meaning of the wave function.

First of all, we argue that for a one-body quantum system with mass m and charge Q, the corresponding physical entity described by its wave function, $\psi(x, t)$, is massive and charged, and the effective mass and charge density in each position x is $|\psi(x, t)|^2 m$ and $|\psi(x, t)|^2 Q$, respectively.

---

[3] Note that different from the present argument, the PBR argument does not rely on knowing the state being prepared, and knowing the state being prepared does not help for the PBR argument either.

[4] This is in accordance with the realist view on the relationship between theory and reality, according to which the theoretical terms expressed in the language of mathematics represent the entities existing in the physical world. Moreover, if a realist denies the existence of a physical entity in some region of space where the spatial wave function of the system is not zero, then he or she will need a new entity different from that described by the wave function and a new dynamics different from the Schrödinger equation to explain the result of a local protective measurement made in the region (see the example given below). We will not consider such theories here (see Chapter 7).

The existence of effective mass and charge distributions can be seen from the Schrödinger equation that governs the evolution of the system. The Schrödinger equation for the system in an external electrostatic potential $\varphi(x)$ is

$$i\hbar\frac{\partial\psi(x,t)}{\partial t} = \left[-\frac{\hbar^2}{2m}\nabla^2 + Q\varphi(x)\right]\psi(x,t). \tag{15.1}$$

The electrostatic interaction term $Q\varphi(x)\psi(x,t)$ in the equation indicates that the physical entity described by $\psi(x,t)$ has electrostatic interaction with the external potential in all regions where $\psi(x,t)$ is non-zero. The existence of electrostatic interaction with an external potential in a given region means that there exists electric charge distribution in the region, which can interact with the potential and is responsible for the interaction. Therefore, the physical entity described by $\psi(x,t)$ is charged in all regions where $\psi(x,t)$ is non-zero.[5] In other words, for a charged one-body quantum system, the corresponding physical entity described by its wave function has effective charge distribution in space. Similarly, the existence of effective mass distribution can be seen from the Schrödinger equation for a one-body quantum system in an external gravitational potential:

$$i\hbar\frac{\partial\psi(x,t)}{\partial t} = \left[-\frac{\hbar^2}{2m}\nabla^2 + mV_{\mathrm{G}}(x)\right]\psi(x,t). \tag{15.2}$$

The gravitational interaction term $mV_{\mathrm{G}}(x)\psi(x,t)$ in the equation indicates that the (passive gravitational) mass of the system distributes throughout the whole region where its wave function $\psi(x,t)$ is non-zero. In other words, the physical entity described by the wave function also has effective mass distribution.

The effective mass and charge distributions manifest more directly during a protective measurement, which can measure the actual physical state of a single quantum system. Consider an ideal protective measurement of the charge of a quantum system with charge Q in an infinitesimal spatial region dv around x_n. This is equivalent to measuring the following observable:

$$A = \begin{cases} Q, & \text{if } x_n \in dv, \\ 0, & \text{if } x_n \notin dv. \end{cases} \tag{15.3}$$

During the measurement, the wave function of the measuring system, $\phi(x,t)$, will obey the following Schrödinger equation:

$$i\hbar\frac{\partial\phi(x,t)}{\partial t} = -\frac{\hbar^2}{2M}\nabla^2\phi(x,t) + k\frac{e\cdot|\psi(x_n,t)|^2 dvQ}{|x-x_n|}\phi(x,t), \tag{15.4}$$

---

[5] On the other hand, the existence of effective charge distribution in all regions where $\psi(x,t)$ is non-zero also indicates that there is a physical entity there, which has effective charge distribution in space.

where M and e are the mass and charge of the measuring system, respectively, and k is the Coulomb constant. From this equation, it can be seen that the property of the measured system in the measured position x_n that is able to influence the measuring system is $|\psi(x_n, t)|^2 \mathrm{d}vQ$, the effective charge there.[6] This is also the result of the protective measurement, $\langle A \rangle = |\psi(x_n, t)|^2 \mathrm{d}vQ$. When divided by the volume element, it gives the effective charge density $|\psi(x, t)|^2 Q$.[7]

Now we will analyze the physical origin of the effective charge distribution.[8] What kind of entity or process generates the effective charge distribution in space or the physical efficiency of the quantity $|\psi(x, t)|^2 \mathrm{d}vQ$? It can be expected that the answer will help us understand the ontological meaning of $|\psi(x, t)|^2$ and the wave function itself.

There are two possibilities: the effective charge distribution of a one-body system can be generated by either (1) a continuous charge distribution with density $|\psi(x, t)|^2 Q$ or (2) the motion of a discrete point charge Q which spends time $|\psi(x, t)|^2 \mathrm{d}v\mathrm{d}t$ in the infinitesimal spatial volume $\mathrm{d}v$ around x in the infinitesimal time interval $[t, t + \mathrm{d}t]$.[9] Correspondingly, the underlying physical entity is either a continuous entity or a discrete particle. For the first possibility, the charge distribution exists throughout space at the same time, while for the second possibility,

---

[6] Note that even in standard quantum mechanics, it is also assumed that the above interaction term indicates that there is a charge $|\psi(x_n, t)|^2 \mathrm{d}vQ$ in the region $\mathrm{d}v$. If there exists no effective charge in the measured position which is responsible for the shift of the pointer of the measuring device there, then a new entity existing elsewhere (which is different from the entity described by the wave function) and a new dynamics for the entity (which is different from the Schrödinger equation) will be needed for a realistic explanation of the shift of the pointer. For example, suppose the measured wave function is localized in two widely separated regions and the measurement is made in one region. If there is nothing in the measured region, then the result of the protective measurement made there or the shift of the pointer of the measuring device there can only be explained by the existence of certain entity in other regions via action at a distance. This will require a wholly new theory different from quantum theories, which will not be considered here. Note that in Bohmian mechanics, the Bohmian particles alone cannot fully explain the result of such a protective measurement (cf. Esfeld et al., 2013).

[7] Similarly, we can protectively measure another observable $B = \frac{\hbar}{2mi}(A\nabla + \nabla A)$. The measurements will give the electric flux density $j_Q(x, t) = \frac{\hbar Q}{2mi}(\psi^*\nabla\psi - \psi\nabla\psi^*)$ everywhere in space (Aharonov and Vaidman, 1993).

[8] Historically, the charge density interpretation for electrons was originally suggested by Schrödinger in his fourth paper on wave mechanics (Schrödinger 1926). Schrödinger clearly realized that the charge density cannot be classical because his equation does not include the usual classical interaction between the densities. Presumably since people thought that the charge density could not be measured and also lacked a consistent physical picture, this interpretation was soon rejected and replaced by Born's probability interpretation. Now protective measurements help re-endow the effective charge distribution of an electron with reality. The question is then how to find a consistent physical explanation for it. Our following analysis may be regarded as a further development of Schrödinger's original idea to some extent. For more discussions on Schrödinger's charge density interpretation see Bacciagaluppi and Valentini (2009) and Gao (2013b).

[9] Note that the expectation value of an observable at a given instant such as $\langle A \rangle = |\psi(x_n, t)|^2 \mathrm{d}vQ$ is either the physical property of a quantum system at the precise instant (like the position of a classical particle) or the limit of the time-averaged property of the system at the instant (like the standard velocity of a classical particle). These two interpretations correspond to the above two possibilities. For the latter, the observable assumes an eigenvalue at each instant, and its value spreads all eigenvalues during an infinitesimal time interval. Moreover, the time spent in each eigenvalue is proportional to the modulus squared of the wave function of the system there. In this way, such ergodic motion generates the expectation value of the observable in an infinitesimal time interval (see also Chapter 3). We will discuss later whether this picture of ergodic motion applies to properties other than position.

at every instant there is only a localized, point-like particle with the total charge of the system, and its motion during an infinitesimal time interval forms the effective charge distribution. Concretely speaking, at a particular instant the charge density of the particle in each position is either zero (if the particle is not there) or singular (if the particle is there), while the time average of the density during an infinitesimal time interval around the instant gives the effective charge density. Moreover, the motion of the particle is ergodic in the sense that the integral of the formed charge density in any region is equal to the expectation value of the total charge in the region.

In the following, we argue that the existence of a continuous charge distribution may lead to inconsistency. If the charge distribution is continuous and exists throughout space at the same time, then any two parts of the distribution, like two electrons, will arguably have electrostatic interaction described by the interaction potential term in the Schrödinger equation. However, the existence of such electrostatic self-interaction for a quantum system contradicts the superposition principle of quantum mechanics (at least for microscopic systems such as electrons). Moreover, the existence of the electrostatic self-interaction for the effective charge distribution of an electron is incompatible with experimental observations as well. For example, for the electron in the hydrogen atom, since the potential of the electrostatic self-interaction is of the same order as the Coulomb potential produced by the nucleus, the energy levels of hydrogen atoms would be remarkably different from those predicted by quantum mechanics and confirmed by experiments if there existed such electrostatic self-interaction. By contrast, if there is only a localized particle at every instant, it is understandable that there exists no such electrostatic self-interaction for the effective charge distribution formed by the motion of the particle. This is consistent with the superposition principle of quantum mechanics and experimental observations.

Here is a further clarification of this argument. It can be seen that the non-existence of self-interaction of the charge distribution poses a puzzle. According to quantum mechanics, two charge distributions, such as two electrons which exist in space at the same time, have electrostatic interaction described by the interaction potential term in the Schrödinger equation, but for the effective charge distribution of an electron, any two parts of the distribution have no such electrostatic interaction. Facing this puzzle one may have two choices. The first one is simply admitting that the non-existence of self-interaction of the effective charge distribution is a distinct feature of the laws of quantum mechanics, but insisting that the laws are what they are and no further explanation is needed. However, this choice seems to beg the question and is unsatisfactory in the final analysis. A more reasonable choice is to try to explain this puzzling feature, e.g. by analyzing its relationship with the existing form of the effective charge distribution. The effective charge distribution has two possible existing forms after all. On the one hand, the non-existence of self-interaction of the distribution may help determine which possible form is the actual one. For example, one possible form is inconsistent with this distinct feature, while the other possible form is consistent with it. On the other hand, the actual existent form of the effective charge distribution may also help explain the non-existence of self-interaction of the distribution. This is just what the above argument has done. The analysis establishes a connection between the non-existence of self-interaction of the effective charge distribution and the actual existent form of the distribution. The reason why two wave packets of an electron, each of which has part of the electron's charge in effect, have no electrostatic interaction is that these two wave packets do not exist at the same time, and their effective charges are formed by the motion of a localized particle with the total charge of the electron. Since there is only a localized particle at every instant, it is understandable that there exists no electrostatic self-interaction of the effective charge distribution formed by the motion of the particle. By contrast, if the two wave packets with charges, like two electrons, existed at the same time, then they would also have the same form of electrostatic interaction as that between two electrons.[10]

To sum up, we have argued that for a one-body system, the physical entity described by its wave function is a discrete, localized particle. At every instant there is a particle with the mass and charge of the system, while during an infinitesimal time interval around the instant the ergodic motion of the particle forms the effective mass and charge distributions measurable by protective measurements, and the spending time of the particle around each position in space is proportional to the modulus squared of the wave function of the system there.

15.3.2 Many-body systems

In this section, we analyze many-body systems, and present further arguments supporting the above interpretation of the wave function in terms of particle ontology.

For an N-body system, its wave function is defined in a $3N$-dimensional configuration space. If the wave function describes a continuous entity, then this entity exists in the $3N$-dimensional configuration space. It has density and flux density in the configuration space. This view is usually called wave function realism or configuration space realism (Albert, 1996), and it has at least two problems, the so-called "problem of perception" and the "problem of lacking invariances" (Monton, 2002; Lewis, 2004; Solé, 2013).[11] The first problem is that this view needs to explain the manifest three-dimensional character of our perception. The second

---

[10] Note that this argument does not assume that charges which exist at the same time are classical charges and they have classical interaction. By contrast, the Schrödinger–Newton equation, which was proposed by Diósi (1984) and Penrose (1998), treats the mass distribution of a quantum system as classical.

[11] See Maudlin (2013) for other criticisms of wave function realism.

problem is that the dynamical symmetries of the Schrödinger equation for an N-body system include translations and rotations in three independent spatial dimensions – not $3N$, and this rich structure of configuration space is in want of a reasonable explanation. Similarly to the case of one-body systems, the wave function of an N-body system may also describe a discrete particle moving in the $3N$-dimensional configuration space, and its motion forms the density and flux density in the configuration space. For example, the density $|\psi(x_1, x_2, ..., x_N, t)|^2$ is formed by the motion of the particle which spends time $|\psi(x_1, x_2, ..., x_N, t)|^2 dVdt$ in an infinitesimal volume dV around $(x_1, x_2, ..., x_N)$ in the infinitesimal time interval $[t, t + dt]$. This view is another form of configuration space realism, and it also has the above two problems.

In the following, we argue that what the wave function of an N-body system describes is not a physical entity, either a continuous entity or a discrete particle, in the $3N$-dimensional configuration space, but N physical entities in 3-dimensional space, and these entities are not continuous entities but discrete particles. First of all, in the Schrödinger equation for an N-body system, there are N mass parameters $m_1, m_2, ..., m_N$ (as well as N charge parameters etc.). These parameters are not natural constants, but properties of the system; they may be different for different systems. Moreover, it is arguable that different mass parameters represent the same mass property of different physical entities. If a system has N mass parameters, then it will contain N physical entities. Therefore, an N-body system contains N physical entities, and the wave function of the system describes the state of these physical entities.[12] Next, these N entities exist in 3-dimensional space, not in a $3N$-dimensional configuration space. The reason is that in the Schrödinger equation for an N-body system, each mass parameter m_i is only correlated with each group of three coordinates (x_i, y_i, z_i) of the $3N$ coordinates in configuration space. Third, these N entities cannot be continuous entities, which are completely described by density and flux density. The reason is that the density and flux density of N continuous entities which are defined in 3-dimensional space are not enough to constitute the (entangled) wave function defined in a $3N$-dimensional configuration space.

Therefore, it is arguable that the wave function of an N-body system describes the state of N discrete particles in 3-dimensional space (see also Monton, 2002; Lewis, 2004). Concretely speaking, at a given instant, the positions of these N particles in 3-dimensional space can be represented by a point in a $3N$-dimensional configuration space. During an infinitesimal time interval around the instant, these particles move in the real space, and correspondingly, this point moves in the

---

[12] Note also that the wave function of an N-body system, which lives on a $3N$-dimensional configuration space, is not a complete description of the system (even though one assumes that the configuration space has a rich structure that can group the $3N$ coordinates), as it contains no information about the masses and charges of its N sub-systems. This point seems to have been neglected by most researchers.

configuration space, and its motion, like the above case of a particle in configuration space, forms the density and flux density in the configuration space. This
interpretation of the wave function has no problems of configuration space realism.

It is worth noting that we can also protectively measure the charge density (and
electric flux density) of a many-body system in 3-dimensional space. A protective
measurement of the observable $\sum_{i=1}^{N} A_i$ on an N-body system whose wave function
is $\psi(x_1, x_2, ..., x_N, t)$ yields

$$\sum_{i=1}^{N} \langle A_i \rangle = \sum_{i=1}^{N} \int ... \int Q_i |\psi(x_1, ..., x_{i-1}, x, x_{i+1}, ..., x_N, t)|^2 dv_1...dv_{i-1}dv_{i+1}...dv_N dv,$$

(15.5)

where Q_i is the charge of the i-th subsystem. When divided by the volume element dv, it yields the charge density in space. Moreover, the previous analysis
of electrostatic self-interaction also applies to many-body systems. Like a one-
body system, the effective charge distribution of an N-body system are arguably
generated by the ergodic motion of N charged particles, where the time spent by
particle 1 with charge Q_1 in an infinitesimal spatial volume dv_1 around x_1 and
particle 2 with charge Q_2 in an infinitesimal spatial volume dv_2 around x_2 ... and
particle N with charge Q_N in an infinitesimal spatial volume dv_n around x_N is
$|\psi(x_1, x_2, ..., x_N, t)|^2 dv_1...dv_N dt$ in the infinitesimal time interval $[t, t + dt]$, or equivalently, the spending time of the N particles in an infinitesimal volume dV around
each position $(x_1, x_2, ..., x_N)$ in the $3N$-dimensional configuration space in the
infinitesimal time interval $[t, t+dt]$ is $|\psi(x_1, x_2, ..., x_N, t)|^2 dV dt$. Such ergodic motion
of particles may explain the entanglement between the sub-systems of the many-
body system. Its existence also shows that all dynamical possibilities of a quantum
universe can be properly represented in 3-dimensional space (cf. Albert, 1996).

15.3.3 *Ergodic motion of particles*

Which sort of ergodic motion? This is a further question that needs to be answered.
If the ergodic motion of particles is continuous, then it can only form the effective
mass and charge distributions during a finite time interval around a given instant.[13]
But according to quantum mechanics, the effective mass and charge distributions at
a given instant are required to be formed by the ergodic motion of particles during
an infinitesimal time interval around the instant. Thus it seems that the ergodic
motion of particles cannot be continuous but must be discontinuous. This is at least
what the existing theory says. This conclusion can also be reached by analyzing a

---

[13] For other objections to classical ergodic models see Aharonov and Vaidman (1993) and Aharonov, Anandan
and Vaidman (1993).

specific example. Consider an electron in a superposition of two energy eigenstates in two separate boxes. In this example, even if one assumes that the electron can move with infinite velocity, it cannot *continuously* move from one box to another due to the restriction of box walls. Therefore, any sort of continuous motion cannot generate the effective charge distribution that exists in both boxes.[14]

Since quantum mechanics does not provide further information about the positions of the particles at each instant, the discontinuous motion of particles described by the theory is also essentially random. Moreover, the time spent by the N particles of an N-body system around N positions in 3-dimensional space being proportional to the modulus squared of the wave function of the system there means that the (objective) probability density for the particles to appear in the positions is also proportional to the modulus squared of the wave function there (and for normalized wave functions they are equal). This ensures that the motion of these particles forms the right mass and charge distributions. In addition, from a logical point of view, the N particles as a whole must also have an instantaneous property (as a probabilistic instantaneous condition) which determines the probability density for them to appear in the N positions in space; otherwise the particles would not "know" how frequently they should appear in each group of N positions in space. This property is usually called indeterministic disposition or propensity in the literature.[15]

In conclusion, we have argued that the ergodic motion of the particles of a quantum system that forms its effective mass and charge distributions is discontinuous and random, and the probability density for the particles to appear in every group of positions is equal to the modulus squared of the wave function of the system there.

15.3.4 Interpreting the wave function

According to the above analysis, microscopic particles such as electrons, which are described by quantum mechanics, are indeed particles. Here the concept of particle is used in its usual sense. A particle is a small localized object with mass and charge, and it is only in one position in space at each instant. Moreover, the motion of these particles is not continuous but discontinuous and random in nature. We may say that an electron is a quantum particle in the sense that its motion is not

---

[14] One may object that this is merely an artifact of the idealization of infinite potential. However, even in this ideal situation, the ergodic model should also be able to generate the effective charge distribution by means of some sort of ergodic motion of the electron; otherwise it will be inconsistent with quantum mechanics.

[15] Note that the propensity here denotes single case propensity. In addition, it is worth emphasizing that the propensities possessed by the particles relate to their objective motion, not to the measurements on them. By contrast, according to the existing propensity interpretations of quantum mechanics, the propensities a quantum system has relate only to measurements; a quantum system possesses the propensity to exhibit a particular value of an observable if the observable is measured on the system.

continuous motion as described by classical mechanics, but random discontinuous motion as described by quantum mechanics.

Unlike deterministic continuous motion, the trajectory function $x(t)$ can no longer provide a useful description for random discontinuous motion of a particle. It has been shown that the strict description of random discontinuous motion of a particle can be given based on the measure theory (Gao, 2013b). Loosely speaking, the random discontinuous motion of a particle forms a particle "cloud" extending throughout space during an infinitesimal time interval around a given instant t, and the state of motion of the particle at the instant is represented by the density and flux density of the cloud, denoted by $\rho(x, t)$ and $j(x, t)$, respectively, which satisfy the continuity equation $\partial \rho(x, t)/\partial t + \nabla j(x, t) = 0$. The density of the cloud, $\rho(x, t)$, represents the probability density that the particle appears in position x at instant t, and it satisfies the normalization condition $\int \rho(x, t)\mathrm{d}v = 1$.

As we have argued above, for a charged particle such as an electron the cloud is an electric cloud, and $\rho(x, t)$ and $j(x, t)$, when multiplied by the total charge of the particle, are the (effective) charge density and electric flux density measurable by protective measurements, respectively. Thus we have the following relations:

$$\rho(x, t) = |\psi(x, t)|^2, \tag{15.6}$$

$$j(x, t) = \frac{\hbar}{2mi}[\psi^*(x, t)\nabla\psi(x, t) - \psi(x, t)\nabla\psi^*(x, t)]. \tag{15.7}$$

Correspondingly, the wave function $\psi(x, t)$ can also be uniquely expressed by $\rho(x, t)$ and $j(x, t)$ (except for an overall phase factor). This means that the wave function $\psi(x, t)$ also provides a description of the state of random discontinuous motion of a particle.

The description of the state of motion of a single particle can be extended to the motion of many particles. The extension may explain the multi-dimensionality of the wave function. At a given instant, a quantum system of N particles can be represented by a point in a $3N$-dimensional configuration space. During an infinitesimal time interval around the instant, these particles perform random discontinuous motion in 3-dimensional space, and correspondingly, this point performs random discontinuous motion in the configuration space and forms a cloud there. Then, similarly to the single particle case, the state of the system is represented by the density and flux density of the cloud in the configuration space, $\rho(x_1, x_2, ..., x_N, t)$ and $j(x_1, x_2, ..., x_N, t)$, where the density $\rho(x_1, x_2, ...x_N, t)$ represents the probability density that particle 1 appears in position x_1, particle 2 appears in position x_2, ..., and particle N appears in position x_N.[16] Since these two quantities are defined in

---

[16] When these N particles are independent, the density $\rho(x_1, x_2, ..., x_N, t)$ can be reduced to the direct product of the density for each particle, namely $\rho(x_1, x_2, ..., x_N, t) = \prod_{i=1}^{N} \rho(x_i, t)$.

the $3N$-dimensional configuration space, the many-particle wave function, which is composed of these two quantities, is also defined in the $3N$-dimensional configuration space.

One important point needs to be emphasized here. Since the wave function in quantum mechanics is defined at a given instant, not during an infinitesimal time interval around a given instant, it should be regarded not simply as a description of the state of motion of particles, but more suitably as a description of the dispositional property of the particles that determines their random discontinuous motion at a deeper level. In particular, the modulus squared of the wave function determines the probability density that the particles appear in every possible group of positions in space. By contrast, the density and flux density of the particle cloud in the configuration space, which are defined during an infinitesimal time interval around a given instant, are only a description of the state of the resulting random discontinuous motion of particles, and they are determined by the wave function. In this sense, we may say that the motion of particles is "guided" by their wave function in a probabilistic way.

15.3.5 On momentum, energy and spin

We have been discussing random discontinuous motion of particles in position space. Does the picture of random discontinuous motion exist for other observables such as momentum and energy? Since there are also momentum wave functions etc. in quantum mechanics, it seems tempting to assume that the above interpretation of the wave function in position space also applies to the wave functions in momentum space etc. This means that when a particle is in a superposition of the eigenstates of an observable, it also undergoes random discontinuous motion among the eigenvalues of this observable. For example, a particle in a superposition of momentum eigenstates also undergoes random discontinuous motion among all momentum eigenvalues. At each instant the momentum of the particle is definite, randomly assuming one of the momentum eigenvalues with probability given by the modulus squared of the wave function at this momentum eigenvalue, and during an infinitesimal time interval around the instant the momentum of the particle spreads throughout all momentum eigenvalues.

However, there is also another possibility, namely that the picture of random discontinuous motion exists only for position, while momentum and energy etc. are not instantaneous properties of a particle and they do not undergo random discontinuous change either. There are several reasons supporting this possibility. The first is that our previous arguments for random discontinuous motion of particles apply only to position, not to other observables such as momentum and energy etc. For example, since the interaction Hamiltonian for a many-particle system relates to the positions of these particles, not to their momenta and energies, the previous analysis of electrostatic self-interaction applies only to position. Next, the Kochen–Specker theorem requires that under certain reasonable assumptions only a certain number of observables can be assigned definite values at all times (Kochen and Specker, 1967). This strongly suggests that the picture of random discontinuous motion exist only for a certain number of observables. Moreover, since there are infinitely many observables and these observables arguably have the same status, this may further imply that the picture of random discontinuous motion does not exist for any observable other than position. Lastly, the meaning of observables as Hermitian operators acting on the wave function lies in the corresponding ways to decompose (and also to measure) the same wave function. For example, position and momentum reflect two ways to decompose the same spatial wave function. In this sense, the existence of random discontinuous motion for momentum will be redundant.

Therefore, it seems more reasonable to assume that the picture of random discontinuous motion exists only for position. In this view, the position of a particle is the only instantaneous property of the particle defined at instants (besides its wave function), while momentum and energy are properties relating to the state of motion of the particle (e.g. momentum and energy eigenstates), which is formed by the motion of the particle during an infinitesimal time interval around a given instant.[17] Certainly, when a particle is in a momentum or energy eigenstate, we may still say that the particle has definite momentum or energy, whose value is the corresponding eigenvalue. Moreover, when a particle is in a momentum or energy superposition state and the momentum or energy branches are well separated in space, we may also say that the particle has definite momentum or energy in each separated region.

Finally, we note that spin is a more distinct property. Since the spin of a free particle is always definite along one direction, the spin of the particle does not undergo random discontinuous motion, though a spin eigenstate along one direction can always be decomposed into two different spin eigenstates along another direction. But if the spin state of a particle is entangled with its spatial state due to interaction and the branches of the entangled state are well separated in space, the particle in different branches will have different spin, and it will also undergo random discontinuous motion between these different spin states. This is the situation that usually happens during a spin measurement.

---

[17] Note that the particle position here is different from the position property represented by the position observable in quantum mechanics, and the latter is also a property relating only to the state of motion of the particle such as position eigenstates. In addition, for random discontinuous motion the position of a particle in a position superposed state is indeterminate in the sense of the usual hidden variables, though it does have a definite value at each instant. Another way to see this is to realize that random discontinuous motion of particles alone does not provide a way to solve the measurement problem. For further discussions see Gao (2013b).

15.4 Conclusions

Quantum mechanics is a physical theory about the wave function and its time evolution. There are two main problems in the conceptual foundations of quantum mechanics. The first one concerns the physical meaning of the wave function. The second one is the measurement problem, which concerns the time evolution of the wave function during a measurement. Although the meaning of the wave function should be ranked as the first interpretative problem of quantum mechanics, it has been treated as a marginal problem, especially compared with the measurement problem. There are already several alternatives to quantum mechanics which give basically satisfactory solutions to the measurement problem. However, these theories at their present stages have not yet succeeded in making sense of the wave function.

In this chapter, we propose a new approach for solving the problem of interpreting the wave function, which is to analyze the mass and charge properties of a quantum system. First of all, with the help of protective measurements, we argue that the wave function of a quantum system is a representation of the physical state of the system. The argument does not depend on non-trivial assumptions and also overcomes existing objections to the implications of protective measurements. Next, we further analyze the ontological meaning of the wave function. The key is to realize that the Schrödinger equation, which governs the evolution of a quantum system, contains more information about the system than the wave function of the system, which can help unveil the meaning of the wave function. An important piece of information is the mass and charge properties of the system, which are responsible for the gravitational and electromagnetic interactions between systems. We first analyze the mass and charge distributions of a one-body quantum system. It is argued that the mass and charge of a one-body system such as an electron is distributed throughout space in efficiency, and the effective mass and charge distributions manifest more directly during a series of protective measurements, which indicate that the effective mass and charge density in each position is proportional to the modulus squared of the wave function of the system there. By analyzing the origin of the effective charge distribution, we further argue that the effective mass and charge distributions are formed by the ergodic motion of a localized particle with the total mass and charge of the system. Moreover, the ergodic motion of the particle is discontinuous and random, and the probability density that the particle appears in every position is equal to the modulus squared of its wave function there. We then analyze the mass and charge properties of a many-body system. It is argued that the wave function of an N-body system describes the state of N discrete particles in 3-dimensional space.

Based on these analyses, we propose a new ontological interpretation of the wave function in terms of particle ontology. According to this interpretation, quantum mechanics, like Newtonian mechanics, also deals with the motion of particles in space and time. Microscopic particles such as electrons are still particles, but they move in a discontinuous and random way. The wave function describes the state of random discontinuous motion of particles, and at a deeper level, it represents the dispositional property of the particles that determines their random discontinuous motion. Quantum mechanics, in this way, is essentially a physical theory of the laws of random discontinuous motion of particles. It is a further and also harder question what the precise laws are, e.g. whether the wave function undergoes a stochastic and non-linear collapse evolution.

Acknowledgments

I am very grateful to Dean Rickles, Huw Price, Guido Bacciagaluppi, David Miller, Vincent Lam, and Lev Vaidman for insightful comments and discussions. I am also grateful to Peter Lewis and Maximilian Schlosshauer for helpful criticisms on my previous analysis of the implications of protective measurements. This work is partly supported by the Top Priorities Program (Grant No. Y45001209G) of the Institute for the History of Natural Sciences, Chinese Academy of Sciences.

References

Aharonov, Y., Albert, D. Z. and Vaidmen, L. (1988). How the result of a measurement of a component of the spin of a spin-1/2 particle can turn out to be 100. *Phys. Rev. Lett.* **60**, 1351.

Aharonov, Y., Anandan, J. and Vaidman, L. (1993). Meaning of the wave function. *Phys. Rev. A* **47**, 4616.

Aharonov, Y. and Vaidman, L. (1990). Properties of a quantum system during the time interval between two measurements. *Phys. Rev. A* **41**, 11.

Aharonov, Y. and Vaidman, L. (1993). Measurement of the Schrödinger wave of a single particle. *Phys. Lett. A* **178**, 38.

Albert, D. Z. (1996), Elementary quantum metaphysics. In J. Cushing, A. Fine and S. Goldstein (eds.), *Bohmian Mechanics and Quantum Theory: an Appraisal*. Dordrecht: Kluwer, 277–284.

Anandan, J. (1993). Protective measurement and quantum reality. *Found. Phys. Lett.* **6**, 503–532.

Bacciagaluppi, G. and Valentini, A. (2009). *Quantum Theory at the Crossroads: Reconsidering the 1927 Solvay Conference*. Cambridge: Cambridge University Press.

Bell, J. S. (1990). Against 'measurement'. In A. I. Miller (ed.), *Sixty-Two Years of Uncertainty: Historical Philosophical and Physics Enquiries into the Foundations of Quantum Mechanics*. Berlin: Springer, 17–33.

Belot, G. (2012). Quantum states for primitive ontologists: a case study. *European Journal for Philosophy of Science* **2**, 67–83.

Colbeck, R. and Renner, R. (2012). Is a system's wave function in one-to-one correspondence with its elements of reality? *Phys. Rev. Lett.* **108**, 150402.

Dass, N. D. H. and Qureshi, T. (1999). Critique of protective measurements. *Phys. Rev. A* **59**, 2590.

Dickson, M. (1995). An empirical reply to empiricism: protective measurement opens the door for quantum realism. *Philosophy of Science* **62**, 122.

Diósi, L. (1984). Gravitation and the quantum-mechanical localization of macro-objects. *Phys. Lett. A* **105**, 199–202.

Einstein, A., Podolsky, B. and Rosen, N. (1935). Can quantum-mechanical description of physical reality be considered complete? *Phys. Rev.* **47**, 777.

Esfeld, M., Lazarovici, D., Hubert, M. and Dürr, D. (2013). The ontology of Bohmian mechanics. *British Journal for the Philosophy of Science*. First published online September 19, 2013.

Gao, S. (2011a). The wave function and quantum reality, In A. Khrennikov, G. Jaeger, M. Schlosshauer and G. Weihs (eds.), *Proceedings of the International Conference on Advances in Quantum Theory*, AIP Conference Proceedings 1327, 334–338.

Gao, S. (2011b). Meaning of the wave function, *International Journal of Quantum Chemistry* **111**, 4124–4138.

Gao, S. (2013a). On Uffink's criticism of protective measurements. *Studies in History and Philosophy of Modern Physics* **44**, 513–518.

Gao, S. (2013b). Interpreting quantum mechanics in terms of random discontinuous motion of particles. philsci-archive.pitt.edu/9589/.

Kochen, S. and Specker, E. (1967). The problem of hidden variables in quantum mechanics. *Journal of Mathematics and Mechanics* **17**, 59–87.

Leifer, M. S. and Maroney, O. J. E. (2013). Maximally epistemic interpretations of the quantum state and contextuality. *Phys. Rev. Lett.* **110**, 120401.

Lewis, P. J. (2004). Life in configuration space. *British Journal for the Philosophy of Science* **55**, 713–729.

Lewis, P. G., Jennings, D., Barrett, J. and Rudolph, T. (2012). Distinct quantum states can be compatible with a single state of reality. *Phys. Rev. Lett.* **109**, 150404.

Maudlin, T. (2013). The nature of the quantum state. In A. Ney and D. Albert (eds.), *The Wave Function*, Oxford: Oxford University Press, pp. 126–154.

Monton, B. (2002). Wave function ontology. *Synthese* **130**, 265–277.

Ney, A. and Albert, D. Z. (eds.) (2013). *The Wave Function: Essays on the Metaphysics of Quantum Mechanics*. Oxford: Oxford University Press.

Patra, M. K., Pironio, S. and Massar, S. (2013). No-go theorems for ψ-epistemic models based on a continuity assumption. *Phys. Rev. Lett.* **111**, 090402.

Penrose, R. (1998). Quantum computation, entanglement and state reduction. *Phil. Trans. R. Soc. Lond. A* **356**, 1927.

Pusey, M., Barrett, J. and Rudolph, T. (2012). On the reality of the quantum state. *Nature Phys.* **8**, 475–478.

Schlosshauer, M. and Fine, A. (2012). Implications of the Pusey–Barrett–Rudolph quantum no-go theorem. *Phys. Rev. Lett.* **108**, 260404.

Schlosshauer, M. and Fine, A. (2013). No-go theorem for the composition of quantum systems. *Phys. Rev. Lett.* **112**, 070407.

Schrödinger, E. (1926). Quantizierung als Eigenwertproblem (Vierte Mitteilung). *Ann. d. Phys.* (4) **81**, 109–139. English translation: Quantisation as a problem of proper values. Part IV, Reprint in Schrödinger, E. (1982). *Collected Papers on Wave Mechanics*. New York: Chelsea Publishing Company, pp. 102–123.

Solé, A. (2013). Bohmian mechanics without wave function ontology. *Studies in History and Philosophy of Modern Physics* **44**, 365–378.

Unruh, W. G. (1994). Reality and measurement of the wave function. *Phys. Rev. A* **50**, 882.

Wallden, P. (2013). Distinguishing initial state-vectors from each other in histories formulations and the PBR argument. *Found. Phys.* **43**, 1502–1525.

Index